Build Your Own Home Theater

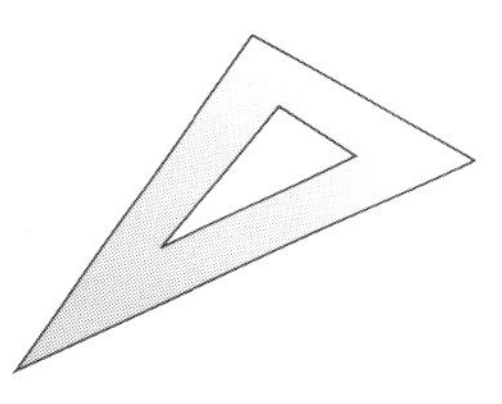

Build Your Own Home Theater

Robert Wolenik

SAMS PUBLISHING

A Division of Prentice Hall Computer Publishing
201 West 103rd Street, Indianapolis, Indiana 46290

BUILD YOUR OWN HOME THEATER • BUILD YO

For Marc, David, and Jason

FIRST EDITION

International Standard Book Number: 0-672-30381-7

Library of Congress Catalog Card Number: 93-85102

96 95 94 93 4 3 2 1

Interpretation of the printing code: the rightmost double-digit number is the year of the book's printing; the rightmost single-digit, the number of the book's printing. For example, a printing code of 93-1 shows that the first printing of the book occurred in 1993.

Composed in 1Stone Serif and MCPdigital by Prentice Hall Computer Publishing

Printed in the United States of America

Trademarks

Publisher
Richard K. Swadley

Associate Publisher
Jordan Gold

Acquisitions Manager
Stacy Hiquet

Acquisitions Editor
Jordan Gold

Development Editor
Rosemarie Graham

Senior Editor
Tad Ringo

Manuscript Editor
Anne Clarke Barrett

Editorial Coordinator
Bill Whitmer

Editorial Assistant
Sharon Cox

Technical Reviewer
Mike Hensley

Marketing Manager
Greg Wiegand

Cover Designer
Tim Amrhein

Director of Production and Manufacturing
Jeff Valler

Imprint Manager
Kelli Widdifield

Book Designer
Michele Laseau

Production Analyst
Mary Beth Wakefield

Proofreading/Indexing Coordinator
Joelynn Gifford

Graphics Image Specialists
Dennis Sheehan
Sue VandeWalle

Production
Nick Anderson
Ayrika Bryant
Lisa Daugherty
Mitzi Foster Gianakos
Dennis Clay Hager
Stephanie McComb
Sean Medlock
Juli Pavey
Linda Quigley
Tonya R. Simpson
Dennis Wesner

Indexer
Suzanne Snyder
John Sleeva

Overview

Contents

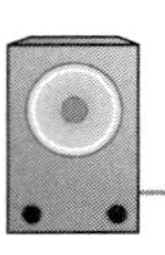

Acknowledgments

The following people contributed to the accuracy and completeness of this work in both large and small ways. Bonnie Mason tirelessly secured photos and permissions. Kimberly Yost and Anthony Grimani of LucasFilm Ltd. spent time helping perfect the material on Home THX. Eileen Tuuri and Roger Dressler of Dolby Labs helped explain Dolby Surround, Dolby Stereo, and Dolby Surround Pro Logic. Stephen Shenefield of Boston Acoustics provided his thoughtful perceptions on the field. David Yaun of Sony Electronics Inc. persisted in obtaining information. Richard Roher of Roher Communications helped secure CEDIA photos. Moe Rubenzahl of Videonics provided insight into new applications for video. Peter Schindo of Panasonic provided marvelous stories of the early days of audio/video (when they used tubes).

About the Author

Robert Wolenik is editor of *Camcorder Magazine* and *ComputerVideo Magazine,* nationally distributed newsstand publications dealing with video cameras and editing, desktop video, and home theater environments. He formerly was editor of *Home Satellite TV* and *Super Television,* consumer electronics publications, as well as managing editor of *COINage Magazine* and *Westerner Magazine.* He has written more than 20 published books in the electronics and computer fields, including *The Camcorder Survival Guide* by SAMS, a series on *Mastering Microsoft Word* by COMPUTE! Books, a series on Epson printers by Reston/Prentice Hall, and *Buyers Guide To Home Computers* by Pinnacle Books.

The author is an avid fan of all types of consumer electronics and is perpetually upgrading his own home theater.

INTRODUCTION

The World of Home Theater

It's an all-new *Star Wars* flick. Or it's a great new mystery thriller. Or it's a romance with that phenomenal new star you like, whose looks and sultry voice have taken the country by storm. Most importantly, it's a movie that promises to be an enthralling sound-and-sight experience.

So you head off to your local cinema, pay seven dollars for a movie ticket, buy the obligatory popcorn, sit down 12 rows from the front, and wait for the action. Unfortunately, most of the action isn't on the screen. The theater quickly fills with noisy teenagers, wailing babies, and gabby couples who (you suspect) will have more to say to each other than the movie will to them. People walk over your feet getting from one side of your row to the other and out of the corner of your eye you watch some rather unsavory-looking individuals cleaning their fingernails with long knives.

Finally the lights dim, and the projector lights up the screen—only instead of the movie you came to see, a dimwitted cartoon welcomes you to the theater. When the cartoon character asks the audience to refrain from smoking or talking, loud boos and jeers erupt.

Then a wild visual kaleidoscope of colors and lines with booming drums and tinkling bells tells you that you're hearing THX, the LucasFilm sound system. It would be magnificent... except for the couple two rows behind arguing about whether they locked their car when they left the parking lot.

Next come previews of coming attractions. Sometimes these are better than the actual films, so you tolerate them. After the fourth one, though, you begin to drum your fingers on the arm rest. Where's the movie?

Finally, it begins. The sound rises, the curtains widen, and the screen fills—only it's out of focus. The projectionist, attempting to correct it, instead overcorrects and it remains ever-so-slightly blurry. Just about this time the person to your right starts coughing. "Sorry," he says, "I god a code."

Oh well, what can you do?

You can do better.

If you love the movie theater, you can create one in your own home—only yours can be better. Your movies will always be in focus. You don't have to limit yourself to films; you can see anything on television as well. The sound will be pure and clear, and without coughs, chatter, or other interruptions, so you can hear it. You can invite just the friends with whom you want to be. And you can fast-forward through previews or cartoons. Best of all, the popcorn will be fresh, with real butter, and won't cost $4.00 for a 25¢ bag!

Enter the Home Theater

Keep in mind that today the rich and famous have their own home theaters. Donald Trump doesn't need to pay seven bucks to see Sylvester Stallone in *Rocky 27*; he can see it at home. Jack Nicholson may sneak into a film preview after the show begins, but you can be sure it's more to see audience reaction than to see the flick, which he's already watched in peace at his own private home theater. Even President Clinton can watch just what he wants in the private viewing theater in the White House.

Check it out. The next time you're driving around on a leisurely Sunday, find some "exclusive" new homes being built nearby you. If they cost lots of money, chances are they include one room that's set up to be a home theater.

We're living in the audio/video age. Those who can, embrace it. It's a sight-and-sound world, and to not participate in it is like being a Roman and never seeing Rome—or never seeing a movie on a big-screen TV off a laserdisc.

Of course, you don't have to be rich (or famous, for that matter) to experience home theater. Home theater is one of the most significant technical achievement of the late twentieth century, but that doesn't mean you want to strangle your pocketbook to acquire one.

The Price

A first-class home theater designed by decorators, acoustical engineers, broadcast specialists, and a home builder can easily cost $50,000 or more. However, if you paid $1.2 million for your home, what's another fifty grand?

On the other hand, if you're willing to spend time trying out different audio and video equipment in electronics stores (and I guarantee that once you know what you're after, it's a blast getting into the great stuff that's out there), building some of the cabinets yourself, laying out the room, lugging all the equipment home, and setting it up, you can have *almost* as good a home theater for $5,000 or less.

That's a big "almost," however. It's important to remember that for another $45,000, besides having someone else do the work, you also get some rather fabulous electronics. Placed side-by-side, you can tell the difference between *top-of-the-line* and *very good*. However, you're not going to place what you get next to *grand majestic*—and by itself, the picture and sound of "very good" will take your breath away.

We'll have more to say about actual costs and what you can get in later chapters. For now, figure on around $5,000 for a good system, and as little as fifteen hundred for a starter set.

Doing It Yourself

The key, of course, is doing it yourself. I've always felt that there were only two good reasons for doing something yourself rather than having someone else do it for you: to save money, and because it's fun. Let's consider these separately.

Reason Number One: You can do it for less. Once you know exactly what you need and how to get it, the amount that can be saved is almost unbelievable. (My own home theater cost far less than five grand.) Of course, the requirement is that you must be handy. But not *very* handy.

After all, what we're talking about is putting together components. You don't have to know how to build a television set. You only have to know which wires in back go to the receiver and which to an antenna, and you can find that explained in a later chapter. You don't have to know how to create cabinets from scratch. You can buy finished cabinet parts (painted, finished, and ready to go) at most local hardware stores and assemble them. You don't even have to know how it all works. All you have to do is place it correctly, connect all the wires and it's done. In theory, someone who doesn't know a screwdriver from a pair of pliers can do it. And even if you're all thumbs, isn't it worth taking a chance if you get to save yourself fifty grand or so? (Remember, this book includes detailed instructions.)

Reason Number Two: It's fun. Let's face it, what we're talking about here are adult toys. Get your kids a train set and they can play for hours, even days, hooking it up and running it. Getting an adult a home theater and is the same thing. (Only we're not supposed to think of it that way. Terms like "investment opportunity" and "helps the kids learn" and "easier on your eyes" creep in. But you and I know what it's really all about, don't we?)

Creating a Master Plan

So, if you're thinking about setting up a home theater (and, presumably, you're serious about it; after all, you invested in the price of this book), where do you begin?

Do not, I repeat, *do not* begin by going out and looking at equipment without reading further. Remember, it's fun checking out what's current on the market. The TVs and the sound systems out there are wonderful—and the salespeople do what salespeople do. Chances are, you'll end up buying something the first time out, only to find out later that it will have to be replaced.

Before you do anything else, stop, take a deep breath, and create a master plan. Figure out what you'll need, *totally*. Before you buy even the first CD, plan your entire system. It won't do to blow your budget on a big-screen TV only to discover that now you don't have enough left for a good receiver. That's sort of like buying only one speaker when you want stereo. It's better to get two less-expensive speakers, or wait until you can afford two better-quality ones.

The name of this game is "system." You want a complete home-theater system. Whether you have $1,500 or $15,000 to spend, you'll want the same type of equipment. The difference will be in the level of performance. You may want to build your system slowly over time adding better components as you get the money. That's fine—but know what you're doing so that you get the right level of equipment from the beginning.

Here, therefore, are the questions you should be asking yourself:

1. What's the total amount of money I can (or am willing) to spend on my home-theater system?
2. Do I have it to spend all at once, or will I build my system over time?
3. How complete do I want my system to be? (For example, will it include a camcorder and a satellite dish?)
4. Where am I going to put it? Do I have a big enough room? How small a room can I get it into?
5. What are the specific components I'll need, and what will I have to do myself?

After you've answered these questions, you're ready to hit the stores. But take your time. And finish reading this book. It will give you all sorts of valuable tips and clues.

CHAPTER 1

What Goes into a Home-Theater System?

My roommates in college thought I was crazy when I put together my first "hi-fidelity" audio system. That was nearly thirty years ago, and when I hauled in that single Jensen monaural speaker system—which was three feet high, four feet wide and two feet thick—they were sure I was ready for the loony bin. Then, with a HeathKit pre-amplifier and "ultralinear" amplifier, which I had built myself to save costs, plus a Girrard turntable (the best ever made up to then—some say ever), and I was ready.

I connected it all, powered it up, put a long-playing record of the Beatles' *Hard Day's Night* on the turntable, and let it rip. For a few moments no one spoke. We all just listened to the amazingly clear sound, cranked up high. Then some girls from the next building over knocked on the door, asking where the party was, and everyone danced into the night, listening to the raspy voice of Ray Charles and the haunting lyrics of Edith Piaf.

It was a great time; for many of us, it was "when the music began." It has continued into today as first audio and now video have become the defining experience of the late twentieth century for most Americans. Along the way, however, the equipment has changed *dramatically*. While I may still long for the bell-clear tones of that Heath/Jensen monaural setup, what I really crave is the wide-screen, surround-sound ambiance of a modern home theater.

Perhaps you're in the same boat. At some point, you purchased audio and video equipment. Now, you want to get better sight and sound—but the limiting factor is your pocketbook. You want the best, but can you afford it?

In this chapter we're going to journey from the simplest to the most sophisticated home-theater setups. Along the way you may see that what you already have will fit into what you want to get. In other cases, to get what you want will mean "buying all new." Our goal here is to see just what's involved in a home theater.

Tip: Unless you're lavishly wealthy, it's usually a mistake to discard all your current equipment and buy all new A/V equipment. It's better to at least try to build on what you have, replacing existing equipment with better quality equipment over time. It not only saves on the pocket book, but it also keeps your spouse or economically concerned relatives from seriously considering having you committed.

What's Your Preference? A Short Questionnaire

Before proceeding, here's a short questionnaire that will help you determine just what it is you're looking for in a home-theater setup. Each of these topics will be explored in the following chapters.

1. Are you building a "ground up" system, or do you want/need to incorporate existing equipment? ____________________________
2. How often will you use your home theater? __________________
3. Do you have other TVs/stereos in your house that can be used without relying on the home-theater setup? _________________

4. Will the room be used exclusively for home theater? If not, will the room's other purpose(s) conflict with or be impacted by the home-theater system? ______________________________

5. Will the sound levels ever get loud enough to disturb people in other rooms of the house, or in other houses? _______________

6. How important is "ease of use" to you (for example, remote control, easy VCR programming, etc.)? ______________________

7. Which family members will operate the home theater? ____

8. How important is the sound and video quality to you? ____

9. Do you want really deep bass sounds? _______________________

10. What's your primary viewing preference: movies, television, interactive CD, or games? _________________________________

How you answer these questions will determine, to a large degree, the type and quality of home theater you build. In this chapter we examine many of the different kinds of setups that are available. Later, we will discuss many other home-theater alternatives.

A Basic Listening System

Most people today who have not yet embarked on the journey to a home theater have two systems in their home. On the one hand, they have a television set, often an inexpensive model, with the tinny, little speakers that came with it. On the other hand, they may have a basic stereo system consisting of a receiver, cassette player, CD player, and speakers (shown in Figure 1.1). Usually, the systems are separate, and we'll consider them separately. Right now we're going to concentrate on the audio system.

The reason for first looking strictly at audio is that some people will simply connect their existing audio system to their TV and say that they now have a home theater. While there's some truth to that, there's also a lot of misconception. The reason is that the considerations for what goes into a listening-only system are significantly different from those that go into a true, integrated audio/video system.

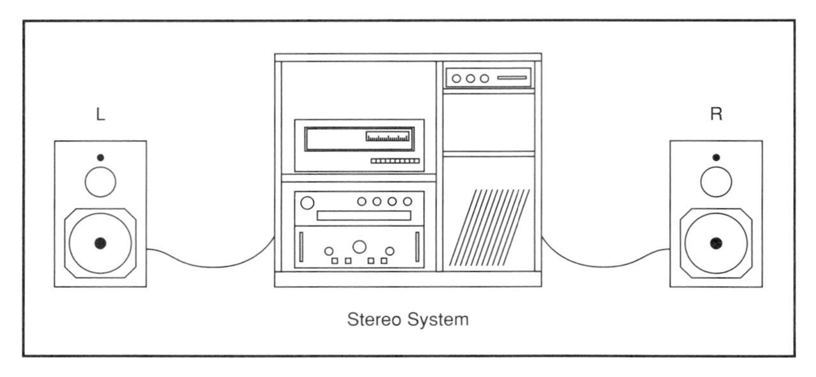

Figure 1.1. A basic stereo listening system.

The guts of a basic home-listening system are usually only a stereo receiver/amplifier and two speakers. The speakers typically are aimed diagonally across a room and the audience (often a single person) sits at the point where each speaker's axis meets and crosses the other. (The *axis* can be thought of as an imaginary centerline, extending perpendicularly from the plane of the speaker's front panel.) This frequently is termed the *stereo sweet spot*, because it's the point where audio from both speakers mix. A person sitting there hears left audio in the left ear and right audio in the right ear, and the brain combines the two into the mixed stereo sound that we've all come to know and love. (This is also known as a *"phantom center channel."* A phantom center channel results when you hear balanced left- and right-audio channels. Your brain processes sounds heard on both channels as originating from midway between the two channels.)

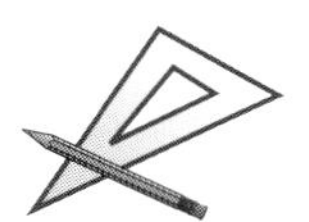

Tip: For just a listening experience, a set of headphones often will work as well as speakers. In fact, for those on a limited budget or in cramped quarters, headphones may be a *preference*.

Listening Systems versus Viewing Systems

The reason this basic audio system works well is that, for listening, we can close our eyes and imagine the instruments and the singers; their location doesn't really matter. As long as the sound envelops us—comes from left and right to create the stereo experience—we're satisfied.

When we add video, things change. With video, our eyes are always open and we're watching characters on a screen. When these characters speak, particularly when we get into wider-screen sets, we expect their voices to come from their location on the screen. If, however, we have only a two-speaker stereo system (particularly if the speakers are far apart), the on-screen voices may come from the left speaker or the right speaker, off-screen. This is a jarring effect, and can momentarily cause our attention to drift from the show. If it happens repeatedly and often, it can make the whole experience distasteful. (This is the reason that many people who first connect their stereo systems to their TVs later disconnect them.)

Thus, much of what's done to improve and modify the audio for home theater works toward creating better *stereo imaging* (how what we hear matches the dramatic placement of what we see on screen), and better synchronization between what happens on screen and what we hear. In the next section we look at how you can optimize stereo imaging with just your existing equipment.

Hooking Up Your Stereo to Your TV

The most obvious problem with a stand-alone television is that, as noted earlier, most inexpensive TV sets have small, tinny speakers. When played at low volume, the big sounds of movies (waves, hoofbeats, gunshots, cars screaming, and so on) are almost totally lost. When we try to compensate for this by cranking up the volume, the sounds often become distorted due to the poor quality of the speakers. Thus, the average, stand-alone TV set is in no way going to be suitable, by itself, for a home-theater experience.

You can improve this by adding your home-stereo system to TV, but in a way that avoids the problem of poor stereo imaging. The easiest way to avoid having people on TV appear to be speaking far off to the left and right when the stereo system is connected is to move your stereo speakers close to the TV. Place them right next to the sides of the TV, as shown in Figure 1.2. You'll lose a significant portion of the stereo separation, but the sound will come from the correct location, and you'll get the much improved sound quality from your audio system added to your TV.

No, this isn't a great solution. But it doesn't cost anything, either!

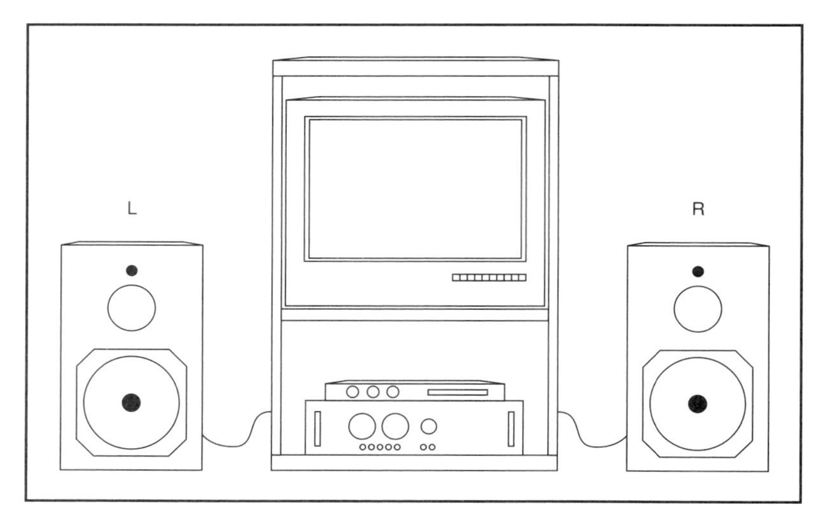

Figure 1.2. A stereo system hooked up to a television for basic A/V.

Caution: Most older (larger) dynamic speakers use big coils and create strong magnetic fields. This stray magnetism can distort the image on your television set by discoloring the picture or causing wavy lines onscreen if older speakers are placed immediately next to it. (Modern speakers usually are shielded.) Therefore, you may have to experiment to get your speakers the right distance from the TV. It's a compromise: close enough for the sound to appear to come naturally from the screen, yet far enough away to avoid causing interference with the picture.

Using Broadcast Sound

To be able to use your existing stereo sound system with your existing TV, you must be able to put TV sound (called *broadcast sound*) through your stereo. There are two ways to handle this:

- You can get the stereo output from your stereo television set.
- You can get a stereo VCR.

Both of these are discussed below.

Using the Stereo from Your TV

First, check to see if your TV has stereo jacks in the back. A quick look will let you know. These are almost always in the form of RCA connectors, pictured in Figure 1.3.

Figure 1.3. RCA jacks on the back of a TV.

If your TV has stereo jacks, you should be able to connect the audio out jacks to a high-level input on your audio receiver/amplifier. (Be aware that in order to use the external audio output on your TV, you may have to flip a switch or change an on-screen control on the TV. Check your user's manual.)

The problem with using your TV set in this manner, however, (assuming it does have audio output jacks) is that you can't get stereo when you play your videotapes. What you need here is a stereo VCR.

Adding a Stereo VCR

Most older (and, especially, less-expensive) VCRs do not offer stereo output. Thus, if there's only one investment that you will make toward getting a better home-theater setup, my suggestion is that you purchase a hi-fi stereo VCR.

While most good VCRs today are reasonably priced, you won't usually find a stereo hi-fi VCR as the inexpensive "loss leader" from a retailer. (A "loss leader" is an item offered at an extraordinarily low price in order to attract customers to the store—where they commonly purchase more items than the one they came in for.)

However, they aren't all *that* expensive—usually only a hundred dollars or so more than monaural VCRs. When you do buy, be sure to purchase one that also offers *multichannel TV sound* (MTS) so that it will decode broadcast stereo.

Caution: The Federal Communications Commission regulates new audio/visual products by requiring that they not make older products unusable. When stereo broadcast sound became available, therefore, a method had to be developed to make such sound useable by older, monaural TVs. Stereo broadcast channels are therefore mixed so that older TVs can accept and transmit them. A feature called *multichannel TV sound* (MTS) uses a "pilot tone" to correctly separate, or *decode*, this mixed signal into left- and right-audio channels.

Drawing the Stereo Off Your VCR

Getting a stereo VCR offers an additional bonus, in that you can hear video tapes in stereo. Instead of drawing the left- and right-stereo sound off the audio outputs of your television set, take them off the audio outputs of your VCR. These then plug directly into a high-level audio input on your stereo AV receiver/amplifier. (See Figure 1.4.)

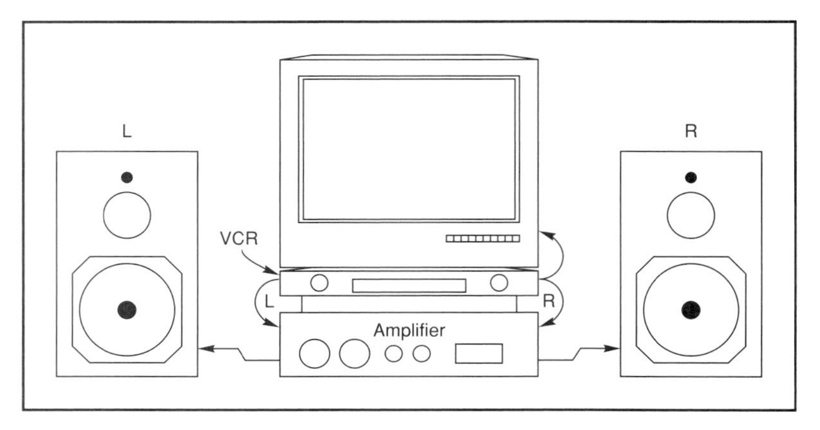

Figure 1.4. Using your VCR for stereo sound.

With the audio coming directly off your stereo VCR, any VHS tape recorded in stereo—which is most of them these days—will yield audio through your A/V receiver/amplifier. In fact, you can get stereo sound from your VCR without even turning on your TV. Regardless of the type of TV set you have, in this kind of setup you're using it only as a monitor.

Further, you actually can get a type of center channel by using the TV's speakers (yes, the tinny things we just talked about not using!). Turn your TV sound down low, so it doesn't overpower the left and right speakers but still offers some center channel sound. This should allow you to move your stereo speakers a bit farther away to the sides for greater stereo separation. The TV speaker won't match the quality of your left- and right-audio speakers. In effect, however, you can create a "poor man's home theater" this way.

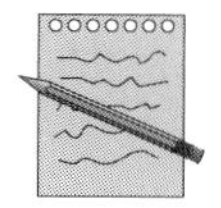

Note: This only works if you use the composite—not RCA jack—connection. The RCA, yellow, connector only contains a video signal. Keep in mind when making connections that different manufacturers may or may not choose to follow industry conventions regarding connection colors. Typically, the connection for the right-audio channel is red, the connection for the left-audio channel is white, and the connection for the video channel is yellow—but this may not be so with your particular equipment. It doesn't matter, however, as long as the outputs match the inputs.

One drawback to this system is that, if you want to tape one program while watching another and your audio is fed from your VCR to your receiver/amplifier, you won't be able to listen to the program you're watching in stereo. In this case, you would want to connect your audio directly from your TV to your stereo, assuming your TV has audio output jacks.

Adding Surround Sound

What we've done so far is figure out a way to utilize equipment you may already have. However, in doing so we've made all sorts of compromises. The sound and sight you see will be a far cry from what's available in a true home-theater setup. Let's now discuss the type of equipment that really will make a difference, again starting with sound.

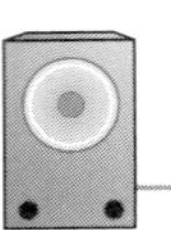

The most dramatic advancement in home stereo in the last ten years has been *surround sound* (often called simply *surround*, and discussed in greater detail in Chapter 5, "Dolby Surround and Dolby Surround Pro-Logic"). As those who have a surround system know, it involves adding two additional speakers behind the listener(s). The back speakers carry slightly different information than the front speakers. Separate wires go to each of the rear speakers so, theoretically, they are two separate channels. However, the signal sent to them comes from the same channel, so they actually are monaural. Both rear speakers produce essentially the same sound.

The first surround-sound systems were built independently by many of the competing manufacturers in the field, and offered a wide variety of methods of accomplishing the same thing: sound that was both slightly out-of-phase with the front speakers and that contained information not found on the front speakers. For example, the front speakers might be creating the sound of airplane propellers while the back speakers might contain some of that sound, slightly out-of-phase (delayed) to allow for the time it takes for the noise to pass by you. They also might contain separate sounds, such as the noise of a tin shed rattling behind you because of the propeller vibrations. In short, the listener would be "surrounded" by sound. The illusion of being there (particularly when you saw a plane's prop turning before you onscreen with a camera pan to a tin shed behind) was greatly enhanced.

Soon after the introduction of surround sound, however, it became apparent that one surround-sound system excelled above all others: Dolby Surround. The reason for this is fairly obvious. The primary surround system used in movie theaters and encoded on movie film is Dolby Stereo. That same system is encoded right along with the movies when they're transferred to videotapes and laser discs. Dolby Surround home systems are set up to correctly decode and play back this information. Here's how this works:

- Part of what makes the Dolby Surround system work better in a movie theater is that it incorporates a specific delay for the rear speakers: the time it takes for sound to travel from the front of the theater to the back. (While this travel time is quite brief in a home-theater space, it still exists.)
- Dolby Surround also incorporates specific *roll-off*, or blocking, of high and low frequencies (that is, high and low sounds) from the surround speakers. Most of these high and low sounds are delivered by the left, right, and other speakers (discussed later in this chapter).

Thus, if you have a Dolby Surround decoder, you can play back almost exactly the same sound that you can hear in a movie theater.

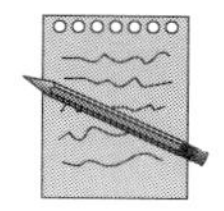

Note: It's important to understand that Dolby Laboratories does not manufacture audio equipment. Rather, it licenses its circuitry for use by other hardware manufacturers. However, Dolby Surround is popular, and is what you very likely (though not necessarily) will get when you buy a surround-sound receiver/amplifier. To be sure, check to see if the unit says "Dolby Surround" on the face. Dolby Laboratories evaluates each product prior to licensing to ensure compliance to their standards.

New Equipment You'll Need

In order to convert your present stereo A/V system to Dolby Surround, you basically need the addition of a Dolby Surround decoder and rear speakers. However, the price of A/V receiver/amplifiers with built-in Dolby Surround has come down so far that you probably can buy a new receiver/amplifier with built-in Dolby Surround for about the cost of a decoder alone.

Also, the rear speakers—which don't need to carry very high or low frequencies—do not have to be of extremely high quality. Thus, you often can get them for a couple of hundred dollars or less.

The Surround-Sound Setup

Up until now, we've basically only taken equipment into account. However, as soon as we move into Dolby Surround, the location of the speakers, and even the setup of the room, becomes important.

The placement of front speakers in a surround-sound setting is similar to that of the standard stereo setup described earlier in this chapter. The placement of the rear speakers is not critical as long as they're generally on either side of and behind the listener so the sound is enveloping. These speakers, which are monaural, are essentially nondirectional and are used only to increase the sound ambiance. In a typical home-theater setup they may be suspended from the ceiling, placed on the floor, or inserted midlevel into cabinets.

It's important that the listener not be distracted by the rear speakers. The dialogue and "important" information is supposed to come from the front, not the back. Therefore, to optimize the system, it's best if the rear speakers aren't seen. This keeps the listener's attention away from them. (Be careful, though, not to hide the speakers to such a degree that sound coming from them is muffled.)

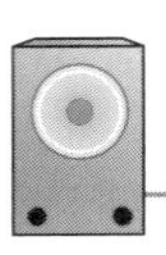

Adding Dolby Surround Pro-Logic

If I've whetted your appetite for surround sound, be aware that (as audiophiles know!) the entire field has taken a great leap forward. It's called Dolby Surround Pro-Logic, and it's an advanced and superior form of surround. The difference in cost is minimal, but the difference in sound is significant.

Thus far in our discussion we've assumed that there is only one listener, and that he or she is sitting at the crossing of the axis of both left- and right-front speakers. At that precise point, the sound will be perfect: a "stereo sweet spot," so to speak. (We may have the TV's speakers on to create a low-grade center channel, but as anyone who has tried it knows, this is a less-than-optimal solution. The results are tinny at best.)

If you think about it for a moment, however, it's rare that only one person watches a movie. In a real theater, there will be people close to the front, way in the back, and far to either the right or left. Even in a home setup there often are several people who sit together to watch a movie. Only one of these, however, can be in the "sweet spot," the perfect location for listening. With the sound systems thus far discussed, those who are off to the side, front, or back will get only a small part of the best sound. In fact, they may not even be able to clearly hear dialogue if it comes, for example, out of the left speaker when they're sitting far to the right.

This problem was recognized early on in movie theaters, and a solution was quickly provided by Dolby Stereo movie theater systems. It was in the form of a separate track that created a true center channel. This center channel feeds to one or more separate speaker(s) placed in the front of the theater. This channel carries the dialogue as well as a combination of left and right information, in phase. Thus, regardless of where you sit in the movie theater, your attention will be focused on the screen.

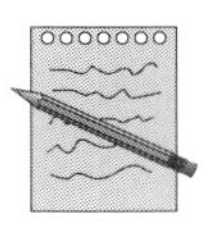

Note: By the way, Dolby Stereo does not simply mean two channels. "Stereo" actually stands for "stereotypical" which refers to sound that is realistic, not sound that is duo. In the cinema, Dolby Stereo actually refers to multichannel sound, which often includes as many as five channels or more.

The same effect is created at home with Dolby Surround Pro-Logic. Figure 1.5 shows a typical Dolby Pro-Logic Surround Sound home configuration. Here you continue to have left- and right-front speakers, which perform the same functions as before. However, added is a center channel speaker placed directly above or below the television set. This center channel speaker carries sounds that are shared by the left and right speakers, as well as most dialogue. (Individual sounds sent to either the left or right speaker remain on that speaker, and are not sent to the center speaker.)

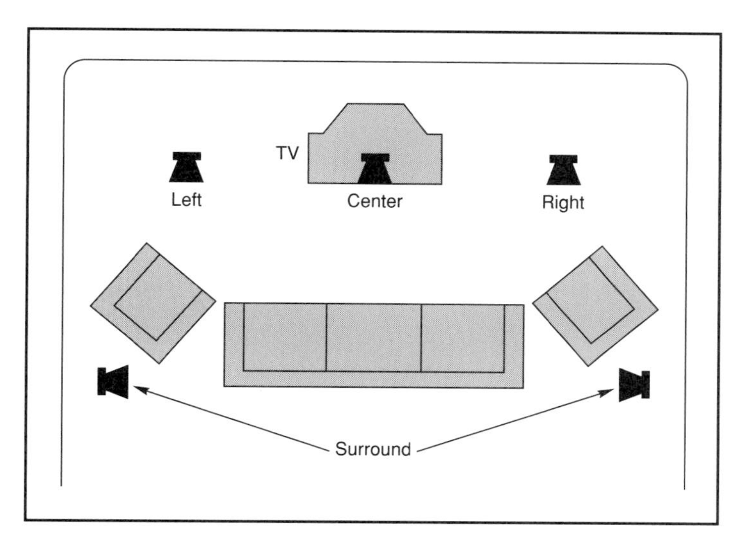

Figure 1.5. A typical Dolby Pro-Logic Surround configuration.

In addition, Pro-Logic offers better *separation*, in terms of phase shifting, between the front and rear speakers than in earlier surround systems. This means that the rear surround speakers add more of the ambient sound to the system at appropriate times.

Thus, for superior sound, you'll want to add Dolby Surround Pro-Logic to your audio system. Again, as with Dolby Surround, this is available as a separate decoder. And again, the price of A/V receiver/amplifiers with built-in Dolby Surround Pro-Logic has come down so far that it's almost better to dump your existing receiver/amplifier and buy a new one. (Besides, the new ones offer lots of great features, such as system-volume controls and switches to handle all of your equipment.)

Adding Speakers for Dolby Surround Pro-Logic

In addition to adding Dolby Surround Pro-Logic, you obviously will need to add a center speaker. While you may keep your existing left- and right-front speakers, this will, unfortunately, be only a poor compromise. The problem is that now *three* front speakers need to be matched. *Matching* ensures that the speakers give the same tonal sound, and is discussed in more detail in the section of this chapter called "Speaker Matching." If the speakers aren't matched, you might be able to hear tonal sound differences between them. The change as the sound moves across the front from left to right or from right to left might be noticeable, and disturbing. (This applies only to frequencies above 100 Hz, as explained in the next section.)

Therefore, if you really want to get a better home-theater system using Dolby Surround Pro-Logic, be prepared to buy at least three new, matched speakers for the front. (And if you can't get three speakers of the same model, at least try to get all three from the same manufacturer since similar materials and production techniques may have been used and the resulting sound may be similar.)

Of course, you don't absolutely *need* to make this purchase. But you'll get much better sound if you do.

Pro-Logic Modes

Pro-Logic decoders or A/V receiver/amplifiers offer several *modes*, which can be additional bonuses depending on the kind of listening you prefer. For example, in *normal mode*, all sound below 100 Hz is fed only to the left and right speakers (not to the center speaker). (100 Hz is 100 cycles per second—the human ear hears low sounds down to about 20 cycles.) Because these low sounds are nondirectional, it's not necessary that they emanate from the center front. They can just as easily be handled by the other speakers. Thus, you need only match the center speaker in terms of higher frequency sound. In other words, you can save a few bucks on the center speaker because it won't need as good a *woofer* (the special, low-frequency element inside the speaker box).

Another plus is the ability to configure your Dolby Surround Pro-Logic system merely for listening enjoyment. While the system is designed for a home theater (that is, for adding sound to movies), in *phantom mode* the information to the center speaker can be eliminated and instead sent to the left and right speakers. While this is useful if you don't have a center speaker, it's even more so if you prefer to turn off the TV and listen to music in the traditional stereo manner.

In *3 Stereo mode*, sound is cut off to the rear surround speakers and their information is instead fed to the center, right, and left speakers. Again, this is useful if you do not have surround speakers, but more so if you do have a center speaker or wish to listen to music in a traditional manner.

In *wide mode*, the center speaker carries the same low sounds (under 100 Hz) as the left and right speakers. Wide mode is useful when you have a full range speaker in the center.

Adding Home THX

The very latest development coming to home theaters is Lucasfilm Home THX (discussed in detail in Chapter 6, "Home THX: Maximizing the Audio"). If you've been to the movie theaters in the past ten years, you've undoubtedly experienced a THX sound system. Introduced in 1983, it's been used in increasing numbers of theaters (currently about 600) in the United States and worldwide. A home version of the system was introduced in 1992, and dozens of manufacturers have been licensed to produce it.

The THX system was developed by Tomlinson Holman, Lucasfilm's technical director—hence the name T(omlinson) H(olman) (e)X(periment). (It also was named for George Lucas's first film, *THX-1138*.) The goal behind the development was to create a system for theaters that accurately reproduced the sounds in George Lucas's films, such as the *Star Wars* trilogy. Now it's possible to get similar sound effects in your home theater with Home THX.

It's important to understand that Home THX is not produced by Lucasfilm (just as Dolby Surround and Dolby Surround Pro-Logic are not produced by Dolby Laboratories). Rather, the company licenses the production of components to manufacturers who agree to meet strict guidelines.

Further, Dolby Surround Pro-Logic is the heart of Home THX. THX just refines it a bit. Here's how:

The Home THX system physically consists of a controller or processor and six speakers (left, center, right, subwoofer, and two surround), each of which must have the THX logo in order to qualify as having met with the strict requirements of Lucasfilm. (To have THX in a movie theater, the theater's system must be inspected by Lucasfilm after installation and several times a year thereafter to be sure it's meeting strict standards.) Home THX essentially adds these four elements:

- Better front-speaker sound
- Less localization for surround speakers
- Speaker matching
- A subwoofer

These elements are discussed next.

How Home THX Creates Better Front-Speaker Sound

THX decodes the audio track signal using Dolby's Pro-Logic Surround Sound techniques. In the process, THX "re-equalizes" the sound for the front channels. In movie theaters, the sound is skewed toward the higher frequencies; in a small room it would sound overly "hissy" or *brilliant*. Home THX restores the original, flat-response characteristics the sound had before the signals were amplified and sent to the front speakers.

In addition, the front speakers are designed to have a wide *horizontal dispersion* as well as focused *vertical directivity*. The result is that the sound from the front is broader, yet the dialogue is cleaner. (These terms are discussed more fully in Chapter 6, "Home THX: Maximizing the Audio.")

Placing Surround Speakers

Like Dolby Surround, Home THX recognizes that the rear surround speakers should have no noticeable *localization*: the listener should hear the sound, but not be able to sense it coming from a localized position. To accomplish this, Home THX *decorrelates* the sound for the rear speakers: although both rear speakers are essentially monaural, THX splits the single surround channel into two uncorrelated outputs for the left- and right-surround speakers.

Further, the surround speakers themselves are *dipolar*, meaning they fire both to the front and back. Also, they're placed at the back and side of the room with the "null" sound of both speakers aimed at the audience. See Figure 1.6 for a view of how the audio surrounds the audience.

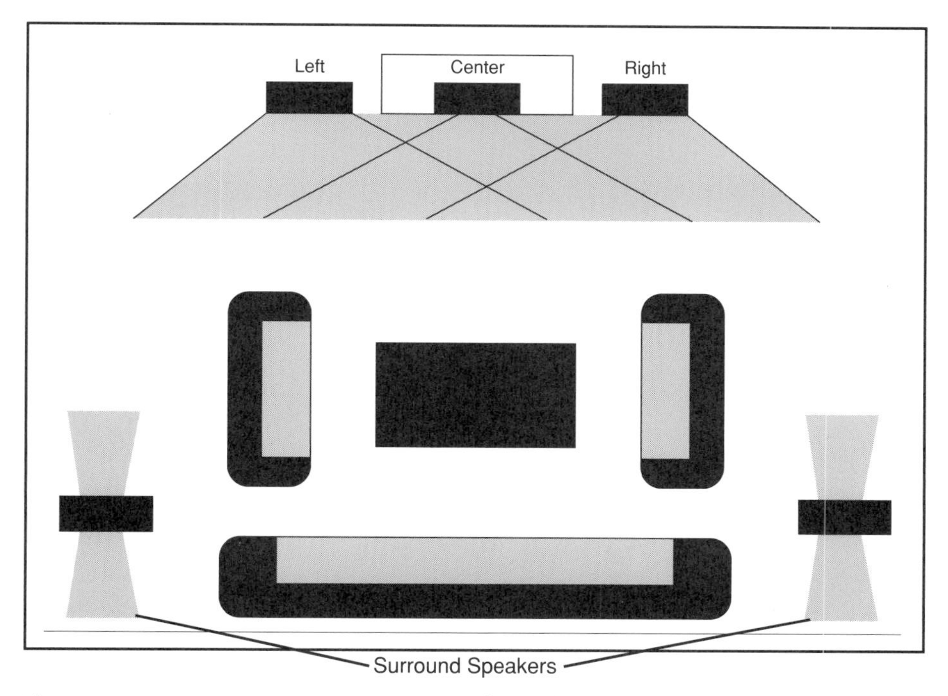

Figure 1.6. A Home THX system, as seen from above.

Speaker Matching

The front and back speakers are electronically *timbre matched*. Tomlinson Holman recognized early on that the human ear distinguishes different sounds largely on the basis of *overtones*, or higher-frequency reverberations. For example, a bass and a piano may strike a note at the exact same frequency (say, 440 Hz for middle A). Yet, the human ear can immediately tell the difference between the sound from a bass and that from a piano based on the overtones.

Similarly, when an airplane passes overhead on-screen, the sound on the speakers moves from the front to the rear. However, because of the configuration of our ears, we tend to perceive sounds differently when they come from the front (the sound bouncing off our heads and ears) and from the side (where the sound, more or less, goes straight down the ear canal). Speakers, in turn, because they're in different positions relative to our ears, are perceived as having their own characteristic timbre. Sound moving from one speaker to another appears to change in

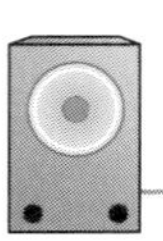

timbre, and the effect is noticeable by the audience. Unless both front and rear speakers are timbre matched, the human ear will instantly detect a difference as soon as the sound is picked up by the rear speakers.

Hence, to avoid this distraction and to ensure that there's no change in timbre as a sound moves from the front sound areas to the surround-sound areas, the Home THX system filters the signals as they cross from front speakers to the surround speakers, and reduces the perception of a change in timbre. (This is done electronically. It would be impossible to physically match speakers perfectly in terms of timbre.)

You'll find much more information on this in Chapter 7, "Speaking of Speakers."

Using a Subwoofer

A *subwoofer* is used to improve deep sounds, allowing you to use smaller front-channel speakers. This also helps in the placement of front speakers in a smaller room. It works like this: instead of feeding the low sounds to the front speakers, we feed them to a single subwoofer. A subwoofer is a larger speaker that carries all *very* low-frequency sound (such as machinery operation, earthquakes, animal growls, and so on). Because such very low-frequency sound is totally nondirectional, the location of the subwoofer is not critical. The subwoofer can be placed on the floor anywhere in the room.

The use of a subwoofer also means that the left and right speakers do not have to carry as wide a range of sound, and can cut off at around 100 cycles. That means that you can get by with a somewhat less-expensive front speaker, built to concentrate more on mid- and high-range sound. Unfortunately, however, the subwoofer itself tends to be pricey—often in the range of $600 or more. Often, a subwoofer comes with its own amplifier, so the cost is understandable.

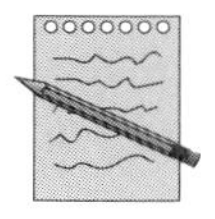

> **Note:** Some Dolby Surround Pro-Logic amplifier/receivers offer subwoofer circuitry unassociated with Home THX. The use of the subwoofer to reduce the load on the front speakers is becoming increasingly popular.

If all of this seems overwhelming, don't panic. In Chapter 6 ("Home THX: Maximizing the Audio"), it's explained in far greater detail. What's important to understand now is that moving up from your present system to Home THX could

require a significant investment. This is because, in addition to the electronic controller, you're also probably going to want to buy new speakers. Purchasing new Home THX speakers is not required and even with your old speakers you should get improved sound. But for the full effect, you would need to purchase all new Home THX speakers: matched speakers in the front and special, dipolar surround speakers in the back.

Buying a Home THX System

Virtually all experts agree that the Home THX sound system is the ultimate audio/video listening experience. Unfortunately, it's also the most pricey. As noted in the last section, you may not want to incorporate all of your existing equipment into Home THX. The least-expensive complete Home THX system currently available offers speakers from Altec Lansing and a controller/processor with a tuner/amplifier from Kenwood for around $5,000 retail. Remember: in order to get Home THX, you must buy products that are licensed by Home THX and show the Home THX logo.

For more information, see Chapter 6, "Home THX: Maximizing the Audio."

The Home-Theater Television Set

Thus far we've been talking almost exclusively about the audio portion of a home theater, mainly because it tends to be the area where the most revolutionary developments have occurred. That doesn't mean, however, that we can simply forget the video portion. For a true home-theater experience you're going to want "bigger and clearer."

At one time everyone in America was thrilled with a black-and-white, 13-inch television set. Today, anything short of a 30-inch color TV with special effects is considered downsized—and for home-theater viewing the sizes often start at 35 inches (tube-type as well as projection-type) and go up from there.

There's nothing to prevent you from using your standard 19-inch TV set with your Dolby Surround Pro-Logic or Home THX system. There will merely be a little something lacking from the true theater experience. Most people want to replace their current small set with one of the newer, better ones. But what essentials do you need to look for?

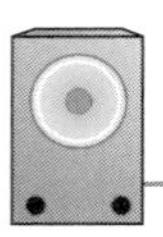

TV in Transition

Modern NTSC sets are the clearest, brightest, and biggest ever built. (NTSC stands for National Television System Committee, the body that standardized the television system in use in North America and Japan.) However, at this writing, the nature of broadcast TV is in transition. The biggest consumer electronics companies have joined together to create a new standard for *high-definition television* (HDTV), which should be available in a few years. HDTV means that the picture you see will have far greater resolution than today's sets offer. In addition, they will have a new shape—wider and shorter. It's called a *16:9 aspect*: 16 inches wide for every 9 inches high. (By contrast, today's conventional sets are 4:3, or four inches wide for every three inches in height.)

Does that mean that you should endure your current small set until the new HDTV sets arrive? Not necessarily.

An interesting study was done a year or so ago (by a major television manufacturer who declines to be named) in which consumers were shown three types of television sets. One was the standard 4:3 set common today. Another was the new 16:9 aspect (wide screen) using today's broadcast standard picture. The last was a 16:9 wide screen using an HDTV signal.

While the vast majority of viewers preferred the 16:9 aspect screen, they didn't notice much difference between HDTV and standard broadcast! They were more concerned with the program that was being broadcast than with improved clarity.

As a result, if you want to buy a TV for your home-theater system, consider some of the wide screen TVs being offered by RCA (ProScan), Panasonic, Mitsubishi, and others.

Otherwise, you may simply want to try one of the standard 4:3 aspect TVs in a larger format. There are tube sets that offer 35-inch and larger screens. And rear projection today actually is brighter, in many cases, than direct view (tube-type) sets!

If you're still confused, check into Chapter 4, "The Big Screen," where we go into detail on big screen TVs. Remember, however, that you're after a home-theater *system*, and the television is only one aspect of that.

Adding Peripheral Equipment

As you build your home theater, there are many other pieces of equipment that you may want to get. These will enhance your enjoyment by improving the quality of what you see and hear. Following are a number you will want to consider:

Adding a Super-VHS VCR

A Super-VHS (S-VHS) VCR, shown in Figure 1.7, is more costly than a standard VHS. However, its resolution is 400 lines or better, while that of a standard VHS VCR is roughly 230 lines. In other words, you get a significantly better picture.

Figure 1.7. VCR with Super-VHS (S-VHS) capabilities.

You can use an S-VHS VCR to tape programs off broadcast TV for later playback. The playback will be superior to that found on a standard VCR. In addition, if you add a camcorder for taking your own pictures, you will get better playback with this format.

Adding S-VHS to your VCR also will add a few hundred dollars or so to the cost. But the results may be well worth it.

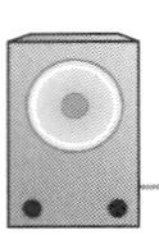

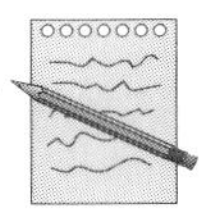

Note: Most S-VHS VCRs come with hi-fi stereo sound built in, so you're getting two features at the same time.

Adding a Laserdisc Player

Laserdiscs, the large video CDs which hold movies, haven't been popular with most of the public. They are, however, very popular with home-theater users! If you're after quality, you should consider adding a laserdisc player to your system.

The reason for this is that the picture and sound quality is significantly improved with a laserdisc player. You have digital sound of the sort you get with a CD player. (Most laserdisc players also can play audio CDs.) Figure 1.8 shows a full-featured laserdisc player.

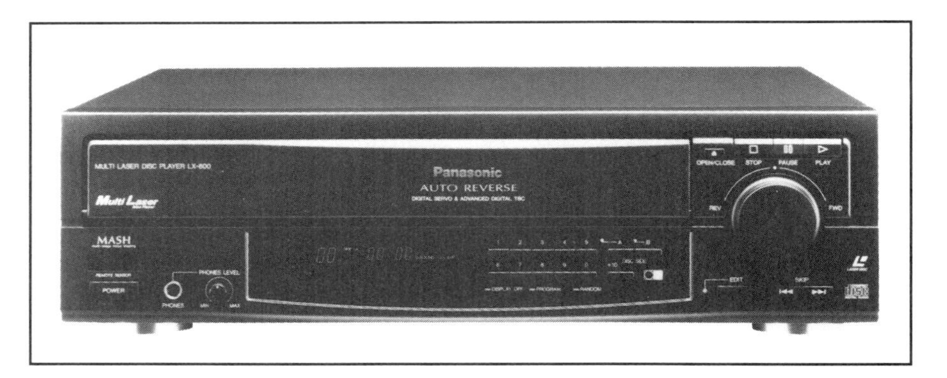

Figure 1.8. A modern, full-featured laserdisc player provides 400 lines of resolution for a retail price of about $500.

The biggest concern here has been with regard to available titles. However, as the demand for laserdiscs increases, so too will the number of titles available. Currently, almost every recent major motion picture is available on laserdisc. Many video rental stores are beginning to carry laserdiscs, as well.

Adding CD-I

Interactive laserdisc players were introduced simultaneously by Philips and Commodore. While Commodore's CDTV entry seems to have vanished from the

marketplace, Philips' Compact Disc—Interactive (CD-I) is benefitting from nearly a billion dollars of marketing and engineering money. Philips CD-I is shown in Figure 1.9.

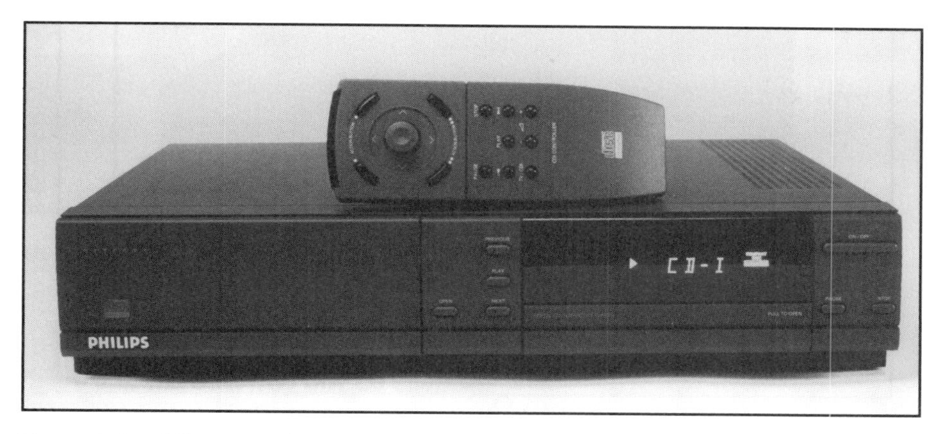

Figure 1.9. Philips CD-I for movies, games, and entertainment.

Thus far, CD-I primarily is used for education and games. The educational aspect is heralded by Compton's Encyclopedia, which brings written, audio, and visual presentations immediately to you on almost any subject. The best known game is golf, which allows you to hit the ball and see it fly over the course to land on the green. Where this differs from other golf games is that the visuals are actual video of a golf course, and the ball does go wherever you hit it!

CD-I, or something similar, definitely is the wave of the future, and no home theater would be complete without a system.

Adding 3DO

3DO (which apparently doesn't stand for anything, but just sounded good to the company founders!) is such a new piece of equipment that most people haven't even heard about it. It's a multipurpose machine that might be called a *video processor*. 3DO allows you to zoom the picture, crop out portions you don't want, create special effects onscreen, play games similar to those on CD-I, create art on your TV screen, and perform a host of other functions. Figure 1.10 shows Panasonic's 3DO.

Figure 1.10. Panasonic version of 3DO for digital interactive video.

Like Dolby Surround and Home THX, 3DO does not manufacture a product, but instead licenses its technology. By 1994, the promise is that companies as widely diverse as Panasonic and AT&T may be coming out with hardware and software for 3DO.

Other Games

A home theater is not complete without a game system, such as those offered by Nintendo or Sega. For advanced games, a home computer with video output does nicely. (Video output is standard on all Amiga computers, but extra on DOS clones—and extra costly for Macs.) Figure 1.11 shows a Super Nintendo gaming system.

Chapter 12, "Game Playing at Home," has much more information on this topic.

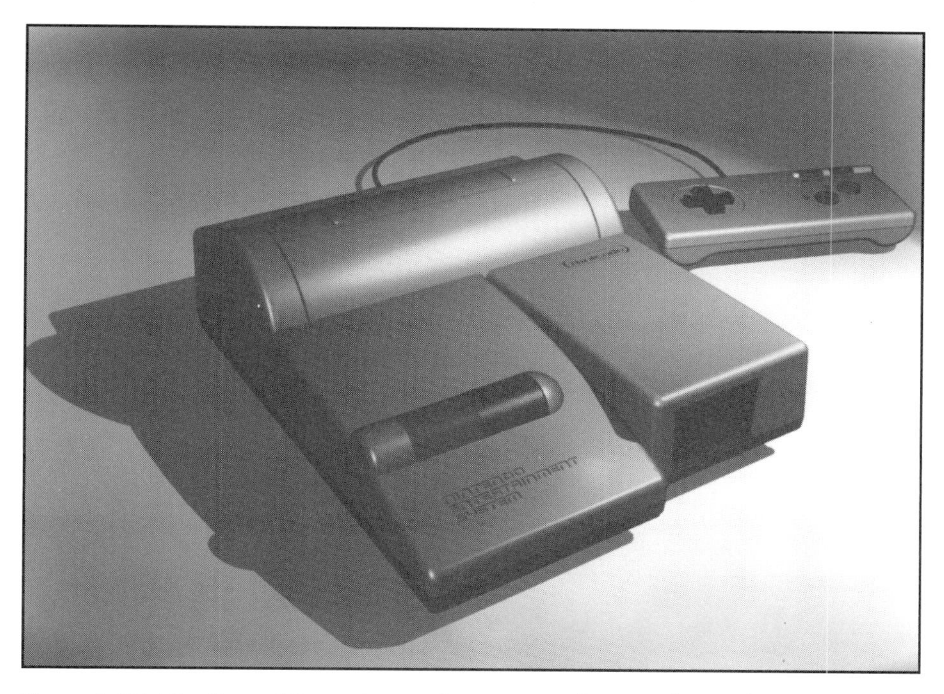

Figure 1.11. The Super Nintendo gaming system plays well on a big screen. (© 1993 Nintendo. Photo courtesy of Nintendo.)

Adding a Satellite Dish

A full home theater doesn't rely on the local cable company to get broadcast television. It picks up broadcast direct from satellites hanging roughly 26,000 miles over the equator. With a *satellite dish* (6 to 12 feet in size and placed in your yard), an *actuator* (to rotate the dish), and a *satellite receiver,* you currently can pick up hundreds of channels.

Be forewarned, however, that many of these channels are scrambled. To get the picture you will need a decoder. These decoders usually are built into the satellite receiver. In addition, once you have the descrambler hardware, you must subscribe to the various channels in order to get the picture; the cost is about the same as for cable. (For example, you could subscribe to HBO, Cinemax, Showtime, and so on.)

The advantage of a satellite dish is a much better picture and sound than your local cable operator or television station can deliver. Also, beginning hopefully in 1994, Hughes, Thomson Electronics, and others in a consortium are building a direct, pay-per-view satellite system. The dish will be only about 18 inches across (because the satellite will use higher-frequency Ku-band technology, requiring smaller dish sizes), and will offer as many as 500 movies to choose from at any one time!

You'll find much more information on this topic in Chapter 13, "Hooking Up to a Satellite."

Adding a Camcorder

While camcorders are not usually considered an essential part of a home theater, with more than two million units sold per year, people increasingly are making their own movies. If you're going to create a home theater, why not become a producer and a director? Figure 1.13 shows a camcorder.

Figure 1.12. Camcorders let you make your own high quality video home movies.

There are many manufacturers of camcorders, and their products are universally good. However, for a home-theater setting you probably will want to stick with one of the higher-quality formats such as Hi8 or Super-VHS.

See Chapter 11, "Camcorders and Home Theater," for more information.

Switchers

One thing that we've touched upon—but should emphasize—is that, as your system grows, you need to be able to switch between different devices such as a VCR, laserdisc, cable, camcorder, or antenna. You can very quickly have an unwieldy system which, in order to operate fully, requires that you be constantly unplugging wires from one piece of equipment and plugging them into another.

One solution to this problem is a *stand-alone switcher*. This device has all inputs coming into it and all outputs going out of it. It's like the central office of the phone system. Once it's connected, all you need to do is adjust the controls on the front to get any combination you desire. (You may have seen a version of this at a retailer, where the sales staff can switch between any receiver and speaker combination for demonstration purposes.)

There are two problems with stand-alone switchers. First, the inexpensive units tend to degrade the signal, so that you lose a slight amount of audio and video. Because you're going for the best possible sight and sound system, this is undesirable.

The other drawback, as suggested above, is money. To get a switcher that does not degrade the signal significantly, it's necessary to spend a lot of money. If you're already considering buying pricey components, this can be an unwanted added expense.

A different solution, and one that most consumers opt for, is to select a receiver/amplifier that's also a switcher. Most modern units offer a host of inputs and outputs that are switchable from the front. Many are controlled by an infrared remote. Thus, the next time you see someone advertising an A/V receiver/amplifier with three switchable outputs (or six, or nine), you'll know how valuable that ability can be to you.

Don't Forget Furniture

While you want to get the theater experience in your home, you can do without one aspect of that experience: lumpy, broken seats. Plan on setting aside some money for comfortable chairs or couches. Also, if possible, set aside some pleasing visuals for the walls. That way the room will feel comfortable even when the stereo and TV aren't on.

See Chapter 10, "Furniture and Lighting That Make the Difference," for much more information on this subject.

As Large or As Small As You Want

A home theater can either be as small or as large as you want to make it. A fully outfitted room could include the following:

- A large-screen TV
- Stereo, Dolby Surround Pro-Logic, or Home THX
- S-VHS hi-fi stereo VCR
- A laserdisc player
- CD-I, 3DO, Nintendo, computer, or other game and interactive equipment
- A satellite dish (which, of course, would be outside)
- A camcorder
- A/V furniture

You certainly don't need it all. But wouldn't it be fun to have?

In the following chapters we'll examine all of the equipment talked about here in much greater detail, and we'll consider layout as well as price.

2 CHAPTER

Fitting the System In Without Buying a New House

You can create a marvelous home theater in any house... in theory. In practice it's a little more difficult. The reason, of course, is that in most cases you're trying to fit the theater into an existing home, not building the home to suit the theater.

If you were building a house from scratch and including a home theater, you could design the room for maximum visual and acoustical presentation. This would undoubtedly include a very big screen, high ceilings, and a large, rectangular room with sound isolation from the rest of the house.

In our case, however, we're going to assume that you've already got your home. You're going to pick one existing room, or a portion of a room, and will be stuck with whatever size and shape that room is. You may be able to build your audio/video electronics from the ground up, but not the location into which it's going.

Therefore, what should you consider in picking a location and what must you do, if anything, to modify it?

Picking the Room (or Portion of a Room)

Here are some important items to consider when picking a location for your home theater:

The size of the room
Intrusive background noise
Whether noise will disturb others in surrounding parts of the house
Unwanted echoes and vibrations within the room
Whether the room is appropriate to your components

Let's consider each separately.

Room Size

Ideally your home theater would have a room all to itself. Realistically, however, your home theater is more likely to be set up in a den, living room, or spare bedroom. In short, it will be in a multipurpose location. Part of the time the room will be used for home entertainment, and another part of the time it will be used for general entertainment, relaxing, sleep, and so on.

Although you can set up your home entertainment system almost anywhere, in choosing a room I would pick a location that is as large as possible. The bigger the room, the bigger the screen—and the louder the sound levels that will be acceptable. If I have a choice, I'll take the largest of a living room, den, or family room.

Beware, however, of choosing a room mainly because it's where you spend most of your time when in the house. For example, you may have a family room off the kitchen where your current television is located. You may watch the TV while in the kitchen preparing meals, and then relax in front of the tube later on. However, the family room may be small. Also, it may be open on one or more walls to the kitchen or other areas. That could mean you'll get unnatural reverberations or sound echoes when the home theater is in place.

Better that you should select a larger, four-walled living room, for example, and place your home theater in it. Most people rarely use the largest room in their home (typically, the living room) anyway. This will give you an opportunity to take advantage of that often wasted space. (Besides, for the family room or kitchen you could always pick up an inexpensive counter TV for part-time viewing.)

The shape of the room is important. Most theaters are longer than they are wide—and for good reason. More people can see the screen more easily this way. Also, the sound is less likely to react with side walls, thus distracting the viewers. The classic way to design a room is to take the height and multiply that by 1.5 to get the room's width. Then multiply the height by 2.5 to get the room's length. (See Figure 2.1.) These proportions normally will distribute sound *resonances* (sound waves bouncing off walls) evenly, and give you a room with good acoustics. Of course, if you're not building a room, but instead using an existing room, try to get a rectangular shaped room, in which you'll place the TV and audio equipment in the center of the narrower wall.

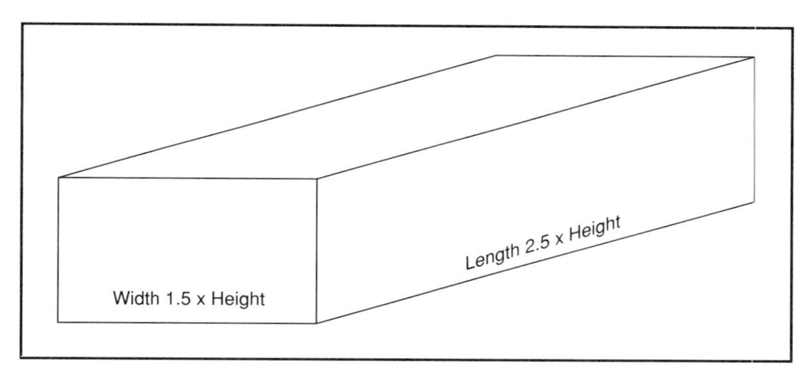

Figure 2.1. Ideal room dimensions are a width that is 1.5 times height and a length that is 2.5 times height.

Higher ceilings are better. Most houses have only eight foot ceilings. However, you'll get better acoustics if your ceiling is ten feet high, as shown in Figure 2.2.

Intrusive Background Noise

Another consideration is *background noise*. This includes rattling from heating or cooling ducts, plumbing pipes, noises from outside on the street, or anything else. When we're living in a house, we often fail to notice these sounds because they occur so regularly. Our ears hear them, but our mind doesn't really pay attention. We tune them out as normal background noise.

Figure 2.2. An ideal home-theater room with good acoustics. (Photo courtesy of CEDIA/Behrens Audio-Video, Jacksonville, Florida.)

However, when you're listening to the *Star Wars* theme and suddenly the music drops to a very quiet level, the sound of that washing machine in the utility room—or the furnace starting up, or even of kids playing ball outside—suddenly can become very loud and annoying.

A good test of any potential room is to go inside, shut all the doors, stand in the middle of the room and then listen carefully for a few minutes for any sounds.

Tip: Listen carefully near wall plugs and light-fixture outlets. If these are connected by metal or other tubing, they often can pass noise in from other rooms. Get your ear close to the heating/ventilation duct as well, and see what noises are coming in from there. Also, check for squeaks, creaks, and groans caused by loose floor boards (or from rooms upstairs when someone is walking). Listen near windows for disturbing street noises.

Some external sounds can easily be eliminated. You can, for example, turn off the washing machine to avoid hearing noise from it while watching your home theater. Other noises, on the other hand, are not so easily removed. For example, you'll want the furnace or air conditioner to continue operating so the room temperature will be comfortable.

For noises that come from electrical outlet boxes or ducts that you cannot eliminate, consider *noise blocking*. Hardware stores sell a foam faceplate gasket for use with outdoor electrical systems. Take off the faceplate (being sure to first turn off the power so you don't get shocked), put on the gasket and replace the cover. It takes less time to do than to explain and should help.

A similar approach can be taken with plumbing connections. Use the foam to seal around any openings.

Ventilation ducts are more of a problem. If the duct opening itself is vibrating (frequently caused by air passing through), consider isolating it by placing some thin foam insulation (sold in hardware stores) around it. This usually just means removing it, wrapping the foam around it, fastening it with duct tape, and replacing it.

On the other hand, if the sounds are being passed through the duct from other rooms, the easier remedy is to identify the room from where the intrusive sound is coming and close the duct in *that* room, at least while you're in your home theater.

If that doesn't work, you may be forced to either buy or build a sound-damping *acoustical baffle*. This essentially means routing the air through a short maze of sound-absorbing material. (Sound damping material often is available from any good audio store. In a pinch, a scrap of heavy carpeting usually will be adequate.) An acoustical baffle, however, is the least-desirable method of damping sound—besides, it usually doesn't work for loud, low noises.

For creaks, squeaks, and groans coming from flooring, the best answer (short of ripping up and replacing the flooring) probably is a *wood screw*. These come in varying lengths up to six inches. Put a wood screw into the noisy joist to permanently squelch the groaning. If you can get to the location of the noise, sometimes adding a wood wedge from the bottom between the floor and joist also will do the job.

For sounds coming in through windows, consider heavy cloth drapes. These absorb sounds very well. For louder, lower noises (which are less directional and tend to carry further), you may want to double-pane the window. It's probably a

mistake to simply buy double-pane glass, as the half-inch or so between panes actually can enhance the sound instead of absorb it. One solution is to install a pane of glass flush with the outside of the wall and another flush with the inside. The usual three-inch difference between panes will help reduce sound transmittal.

Noise That Will Disturb Others

While one consideration is noise disturbing us while viewing our home theater, another consideration is the noise from your sound system disturbing others. With a subwoofer and five other speakers going, the sound in your home theater can be *very* loud. Much of it may escape to adjoining rooms. (The subwoofer, in particular, is going to resonate along the floor and walls.)

The best answer here is to be sure the room selected for the home-theater system is isolated from other rooms in the house, *especially bedrooms*. If this isn't possible, you can insulate the room to prevent sound from leaking out. This, however, often is an expensive undertaking.

Acoustically Insulating a Room

When considering sound-insulating a room, check to see whether only a portion of the room needs to be insulated. Perhaps only one wall adjoins another room in which the home-theater sound will be obtrusive. In that case, only that wall should be insulated. Be aware, however, that low-level sound (under 100 Hz) is nondirectional and often will travel along floor boards into other parts of the house.

A/V stores, hardware stores, and electronics outlets sell a variety of materials that can be used for sound muffling on walls. Most of these are some sort of fiberglass material with a thin, gauze-like exterior. Some of these are attractive and can be nailed or glued directly onto the wall or ceiling you wish to insulate. This often handles most minor noise problems.

Be aware, however, that if you truly want to eliminate noises, particularly low-level growls, the only method guaranteed to work is isolation. This, unfortunately, can be difficult and expensive to handle. Let's first consider floors.

Eliminating Floor Sounds

Most sound transmitted along floor boards comes from direct contact with the sound source, usually, a larger speaker or a subwoofer. To overcome this problem, I suggest you place speakers, especially subwoofers, up off the floor. I recommend at least a 12-inch separation.

Another method is the *floating deck*. This is a specially designed deck that prevents sound waves from traveling directly from the subwoofer enclosure to the floor. You can build this kind of sound-trap easily and inexpensively. It usually consists of a top platform and bottom platform separated by four legs made of some insulating material (such as a tough, white Styrofoam). The subwoofer sounds don't pass through from the top platform to the bottom because of the Styrofoam insulation. Adding just this insulation alone can cure many pass-through sound problems.

Eliminating Wall Sounds

For walls, isolation is more difficult. Most walls are hollow, and the compressed-gypsum Sheetrock (or *wallboard*) on one side can easily transmit sound across the open space to the wallboard on the other. (It's kind of like the sympathetic vibrations an active speaker can induce in an inactive one.)

The easiest solution here (if you don't mind tearing into the wall) is to fill the empty space in the wall, usually with a thick fiberglass lining readily available at most building-supply stores. If this doesn't work, you may want to add a second layer of wallboard to the wall. Because much sound damping simply involves adding density to a wall, the second layer of wallboard should do. (Most inside walls are one-half inch thick and consist of either wallboard or plaster. Adding another half-inch may more than double their sound-absorbing abilities.)

Yet another problem is sound transmittal through the lumber that supports the wall. Noise is carried directly through walls by these *studs*. Wallboard reverberations can be passed directly through the studs to the wallboard in the room on the other side. You can check this effect out yourself by placing your ear next to the wallboard of one room while someone is talking in the next room. As soon as your ear touches the wall, the sound level will jump enormously.

Adding standard fiberglass insulation between wallboards, as noted earlier, will cut down on a large amount of pass-through sound. To eliminate more, you can use USG Acoustical Sealant on the stud surfaces. This will prevent much sound pass-through from wallboard to studs.

However, to truly isolate a wall (or ceiling), it's necessary to create a *double wall*. (See Figure 2.3.) Here, two sets of studs are used, one for each wall, and none of the studs actually touch both walls. If done properly, the double wall will almost entirely eliminate reverberation transmittal. While this is the sort of thing you'd only want to attempt on all walls during construction of a home, you may want to consider it for a single wall that's giving you particularly difficult sound-transmission problems.

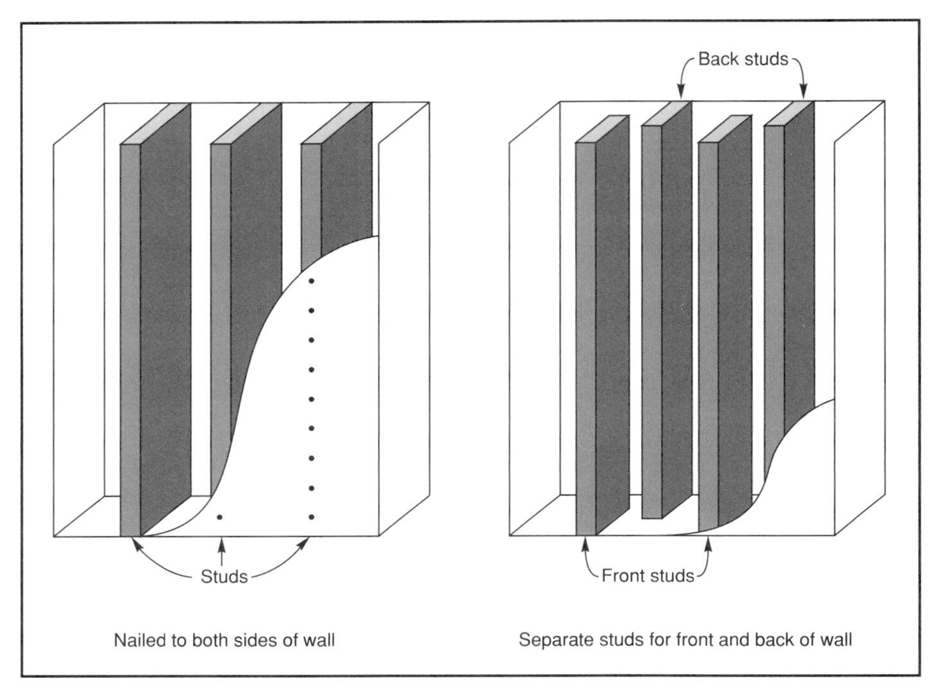

Figure 2.3. A standard wall has wallboard nailed to both sides of the studs. An isolated wall has separate studs for front and back wallboard.

Isolating Ceilings

The easiest way to isolate ceilings is to dump six to ten inches of insulation on the top side of the ceiling. A heavy insulator, such as rock wool, works best. Isolating a problem ceiling may require creating a sound barrier or separate wall in the attic isolating the home-theater room from the rest of the house. Although usually unnecessary, double joints with a Styrofoam-type of insulation between them also may be necessary.

Isolating Doors

For door isolation, you may want to remove an existing hollow-core door and replace with a solid-core unit. Also, sealing the perimeter of the doors on all four sides with a good weatherstripping material will help.

> **Warning:** If you decide to isolate your home theater through construction, be sure to check with your city and county building departments to determine the safety loads required for ceilings, roofs, and support walls.
>
> Also, be sure to check and comply with all local building and electrical codes prior to any modification. Be prepared to obtain permits for most changes.

Ventilation

Fully isolating a room means that you will have effectively cut it off from the rest of the house both in terms of sound and in terms of air. Therefore, you may need to install a vent that will exhaust air from the room either directly outside or to another room that contains the return to your heater/air conditioner. (Fresh air will, presumably, come in through existing ducting but if it doesn't, you may need to install an intake duct as well.) This also will help in eliminating any smells from new carpeting, furniture, electronic equipment and so on. Just be sure that all vents and ducts are sound proofed.

Unwanted Echoes and Vibrations in the Room

Every room will have its own sound characteristics. However, in general, an empty room will have many more reverberations and echoes than a room filled with items that break up the sound waves. Therefore, before you worry too much about unnatural sounds in your room, be sure to fill it with furniture. You may find that once you get your A/V equipment, chairs, couches, rug, wall hangings, and so on in the room, unwanted echoes and vibrations will disappear.

Two kinds of unwanted sound, however, may remain. These may be *echoes*, which are sounds slapping off the walls or ceiling, and *near-wall first reflections*, which are sounds bouncing off the wall closest to the speaker and coming back ever so slightly out-of-phase (causing distortion).

Some echo and vibration problems in a room can be diminished or eliminated simply by adjusting the angle of the speakers. However, if a problem persists with a particular wall, hang one or more *acoustical panels* (fiberglass covered with cloth) at different spots across the room to break up sound dispersion. This often does the trick. Applying acoustical panels to the front and near side walls also may help.

A particularly effective method is to create a cloth-covered fiberglass tube, about a foot or two in diameter, and place it in the room corners.

Yet another method is to use the equalizer on your A/V equipment to balance out the room acoustics. You can increase volume to one speaker while diminishing it to another until a pleasant balance is achieved.

Component Location

Often the way things work out is to pick a room and then say, "Everything's going to fit, period!"

Perhaps it would be better to consider what components you're going to need in your room and then see if the room is appropriate. Maybe you'll want to change rooms... or select different components.

In general, the bigger the room the better. (See optimal room sizes, discussed earlier in this chapter under "Room Size.") One of the most important decisions is the size of the television set. Everyone wants a big set. But, you have to be careful that your set isn't too big for your room.

For example, you may want a 60-inch rear projection TV. If your room is only 12 feet long, however, you're probably going to find that you're forced to sit too close for comfort. Additionally, for the best surround-sound effect, you don't want to be sitting up against the rear wall of the room. You want to be sitting close to the middle, with some space behind you.

Therefore, we want to be sure that our room is big enough to accommodate our list of components. If it isn't, we need a different room... or different components.

Distances from Viewing Screen

How close or how far you should view a TV is largely a matter of personal preference. If you're looking for general guidelines for television sets that are rear-projection units, manufacturers have used a rule of thumb that says the minimum viewing distance is twice the height of the screen. (See Figure 2.4.) However, with recent, super-bright rear-projection screens, some of which are brighter than tube sets, this ratio probably should be increased to around two and a half times the height of the screen. Thus, if you're watching a rear-projection screen that's three feet tall, you wouldn't want to be any closer than about seven or eight feet.

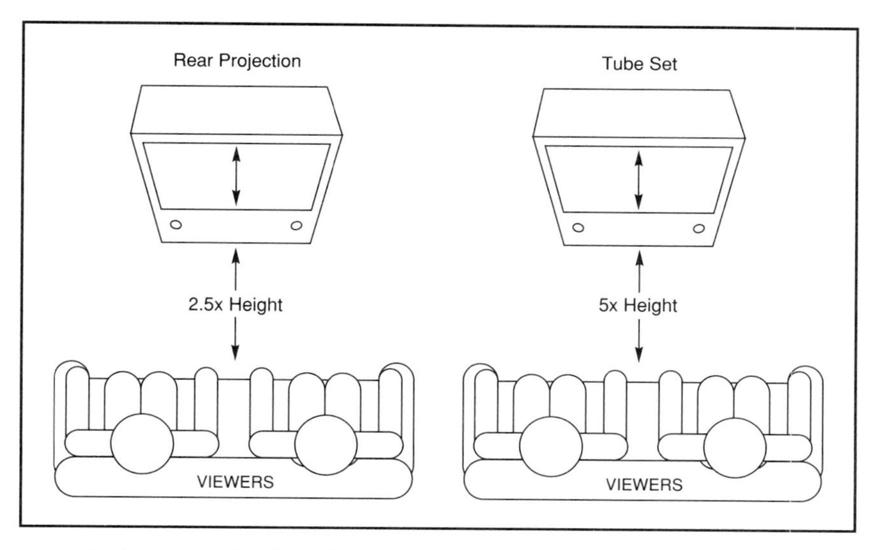

Figure 2.4. Optimum viewing distance varies depending on the type of set being watched.

The problem with sitting closer is that you may find the screen is often fuzzy. In addition, the closer you sit, the more clearly you will see the horizontal matrix that composes the picture. In other words, the physical make-up of the picture itself begins to intrude into your enjoyment of the movie.

For tube-type sets, typically, the minimum viewing distance is considered to be five times the height of the picture. A 20-inch diagonal set is roughly one foot high (using the 4:3 aspect ratio of traditional sets discussed earlier) which means that you would want to sit no closer than about five feet away. Again, I'd prefer a bit further away.

What all of this comes down to is that you need to buy a television set that will allow for easy viewing in the room you have. The proper way to calculate this is to determine the distance from the screen to the viewer, add at least half again that much for the distance from the viewer to the back wall (remember, you don't want to sit right in front of the back wall for good surround sound) and add about two and a half feet for the thickness of the TV. (You usually don't want to recess your TV into a wall for reasons discussed in Chapter 10, "Furniture and Lighting that Make the Difference.")

The following equation may prove helpful in finding the appropriate size of a TV set for any room:

For a Tube Set

1. Measure the length of the room, in feet.
2. Subtract 2.5 feet.
3. Divide the remainder by three.
4. Double the remainder and divide by five. Use the result as the maximum height of the set.

For example, if your room is 14.5 feet in length, subtract 2.5 feet for the depth of the set. Take the remainder of 12 feet and divide by three, yielding four feet. Double that to eight and divide by five, giving a maximum screen height of 1.6 feet.

For a Rear Projection Set

1. Measure the length of the room in feet.
2. Subtract 2.5 feet.
3. Divide the remainder by three.
4. Double the remainder and divide by three. Use the result as the maximum height of the set.

For example, if your room is 14.5 feet in length, subtract 2.5 feet for the depth of the set. Take the remainder of 12 feet and divide by three, yielding four feet. Double that to eight and divide by three, giving a rough maximum screen height of 2.7 feet.

Sound Levels

As far as sound is concerned, most of us only think of how many watts our receiver/amplifier will produce. The other side of the equation, however, is how many watts our speakers require to produce adequate sound. Some systems only achieve true, hi-fidelity sound at very high decibel levels. Others do it at much lower levels.

We need to match our audio system to our room size and what we consider comfortable listening levels. We'll discuss this in far greater detail in chapter seven. Also, we need to be sure that our amplifier can produce the wattage our speakers require just as we need to be sure that our speakers can handle the wattage the amplifier sends out.

In your present house is a room, or a part of a room, that will make a perfectly adequate home theater. You might have to adjust your thinking to a smaller system in order to make it work. However, rest assured: where there's a will, there's a way. If you want to get a home theater in your house, you can.

3 CHAPTER

Building a System for $1,500 to $15,000

How much money do you actually have to spend to build a home-theater system?

Estimates range all over the board. On the one hand, you can buy a speaker system alone for $30,000 or more. On the other hand, you could pay as little as $190 for a 19-inch color television set with on-screen programming and full-function remote control. The real question is how do you accurately determine costs for the type of system you desire?

In this chapter we look at putting together the basic components of a home-theater system. We'll talk about what you get for your money, where savings occur, and what traps to avoid. Hopefully, along the way what can be done (and what can't) for different-sized budgets will become clear.

We do not go into detail on either specialized audio or video considerations in this chapter. These are handled in separate chapters. Here we are mainly concerned with overall system costs.

What Goes into a System

There are basic systems—and then there are complete systems. A basic system includes the following:

Television
Speaker system
AM/FM receiver with Dolby Surround (or Dolby Surround Pro-Logic)
VCR

To make a complete system you would need to add a compact-disc player, a laserdisc player, an interactive machine such as CD-I, and game players such as Super Nintendo or Sega Genesis—not to mention furniture, acoustical baffling, and so on. Here we're only going to look at the basic system. If you want to add to it later on, you can. However, to get up and running, you first need the items listed above.

Building a Basic System for Under $1,500

The biggest challenge in writing this book was to come up with a reasonably good home-theater system for under $1,500. It's easy to come up with a $15,000 system or even one for $5,000. To get good equipment for a low price, however, is very tricky. But it can be done!

It's important to understand that we're not talking about simply putting together a TV set with a stereo. As noted earlier, you can obtain any of a variety of TVs (such as units made for the low-end market by Magnavox, Emerson, Sharp, and others) for around $200. You also can buy a stereo setup, including AM/FM receiver, amplifier, and speakers, for under $200. And you can easily buy a VCR for under $200. Thus, you can put together basic components for as little as $600.

However, this won't really give you the home-theater effect. We're talking here about quality sound *and* sight. And that costs a bit more.

Starter or Permanent System?

The first question you must ask yourself is, "Will this be a permanent system, or is it a starter system?" How you answer this question largely will determine what you buy.

If this is going to be a permanent system, buy the highest quality possible in all areas. Match the components as much as possible.

If this is going to be a starter system, however, sink as much of your money as possible into one component, buying the very best quality you can afford, knowing that this piece is the one you plan to keep for the future. With the rest of your funds, purchase the remaining components for as little as possible, with the understanding that eventually these all will be replaced.

Where to Save Money

It's easier to save money in some areas than in others. We'll consider all of the home-theater components one by one in this section.

The Television Set

To get a TV set today that offers remote control, stereo sound, and on-screen programming with decent speakers and a screen that's at least 30 inches diagonally, you'll spend around $1,000. However, it's possible to save a considerable amount of money by realizing that you don't need most of the functions that add to the price.

For example, you may not need a stereo TV. You're going to get stereo off your A/V receiver or a stereo VCR. You also don't need good speakers on the television (unless you plan to use it when your home-entertainment system is off). Your sound system will have its own speakers, so why spend extra for expensive television speakers?

Further, you don't necessarily need to get a 30-inch or bigger screen. The difference in price between a 27- or 29-inch TV and a 30-inch screen is as much as 40 percent! You pay *40 percent more* for that extra few diagonal inches. If you're willing to settle for a 20-inch screen, you can buy one for about 60 percent less than you can a 30-inch screen.

Of course, the essence of home theater is the visual image. The bigger and better the TV set, the better the home-theater experience. However, if you plan to buy a minimal set now and expand later, why not go with a low cost 20-inch TV? For about $350 you can buy an excellent quality television from Hitachi, Toshiba, RCA, Panasonic or a variety of other vendors in the 20- to 25-inch size. Later, when your pocketbook expands, you can expand the size of your television set along with it.

No matter what kind of a television set you buy, I suggest that you get a full-function remote control and on-screen programming. These are inexpensive features that most manufacturers include, and they add enormously to your enjoyment.

Tip: One of the easiest components to replace is the TV, because it doesn't have to be matched to the rest of your equipment.

Speakers

This usually is the most expensive area, yet the number of manufacturers—from Bose to JBL, from Altec Lansing to Fosgate—is enormous. There are dozens of speaker manufacturers active today. As a result, speakers come in a variety of prices. There is, in fact, no area of the home entertainment system with a wider value range. Unfortunately, the quality of speakers is directly related to the price. You simply aren't going to find inexpensive speakers that sound great. To get good sound costs money.

My personal preference is to spend the majority of my money on speakers. My reasoning is that this is the greatest number of items that must be purchased. After all, you need only one TV and one A/V receiver/amplifier. However, for most systems you need five or six speakers. This means that replacing the speakers later on with better-quality units will be very costly.

What's interesting about speakers is that most (not all) manufacturers offer a variety of price lines. If one manufacturer has a sound you particularly like, you may be able to find a speaker that's smaller, less powerful, and less costly, but sounds similar in a related line. You have a better chance of matching disparate speakers if you stick with one manufacturer who may use the same production techniques and material sources for all its speakers. It's not a guarantee you'll get better matched speakers, but it's a good rule-of-thumb.

In Chapter 7, "Speaking of Speakers," I note that most people feel it's important to have as much power as possible in the speakers. Generally speaking, that means speakers rated at 100 watts or more and capable of producing 100 decibels of sound.

The problem is that there's a tendency for inexpensive speakers to be rated at a very high power, but with significant distortion. Those rated lower often have a cleaner sound. Thus, while power is important, you may be able to sacrifice power for quality and get a better price in the bargain.

The A/V Receiver/Amplifier

A few years ago, amplifier/receiver combinations were amazingly expensive, particularly if they incorporated Dolby Surround or Dolby Surround Pro-Logic. That's changed dramatically with equipment from all major manufacturers recently dropping significantly in price. Good units start at around $300. Top-of-the-line units go for well more than $2,000.

Most manufacturers (such as Sony, Technics, Pioneer, Harman-Kardon, Denon, and others) offer a variety of lines and price ranges. Again, you get what you pay for. The more expensive units offer much cleaner sound, better reception, and more features.

This, however, is an area in which you can save money. A/V receiver/amplifiers have a number of features which, while convenient, may not be vital when you're interested in saving money. (When price is no object, of course, get all the features you can!)

For example, most sets offer memory storage of favorite stations; a very nice, but not necessarily vital, feature. Others offer extensive equalization capabilities. (Not just a bass and treble knob, but control by frequency.) Again, very nice to have, but not absolutely necessary. Yet others offer a large number of mixing capabilities so that you can interchange a variety of inputs and outputs. Check out the number of inputs/outputs you'll need and get a set that's got the right number. Why pay more when you won't need the feature?

Things to look for in an A/V receiver/amplifier include individual balance controls, volume controls for the left, right, center, subwoofer (if you have one), and surround speakers—plus a single volume control that covers all speakers, once they've been set up. Also, you should look for power. The more watts, without distortion, the better.

Unfortunately, just as with speakers, with inexpensive A/V receiver/amplifier models more power often also means more distortion. Check it out by comparing the unit you're considering with other sets. Given pocketbook constraints, I would give up power for quality.

Finally, you want remote control. It's very difficult to set the volume when you're right on top of the set. You need to be able to listen to it from the normal audience position, and you need remote control in order to do this.

The VCR

Home theater means watching movies—and that usually means buying or renting tapes (or laserdiscs). For tapes, you'll need a VCR.

A basic VHS VCR will provide just about as good a video signal as one loaded with features. For the base model, check to be sure it has four *amorphous heads* (an industry standard). If so, you've probably got yourself a solid unit.

For more money, however, you do get some excellent features:

- If you plan to edit movies, add a *flying erase head* for glitch-free editing.
- Add programming for recording television shows when you aren't home.
- Be sure you get on-screen programming—an important feature if you're like me and always have trouble programming the VCR.
- Add Super VHS (S-VHS) for a couple of hundred dollars for superior playback of S-VHS movies. (You can always record your own on an S-VHS camcorder.)
- Add hi-fi stereo if the unit you're considering doesn't come with it (although it usually does). This is about a hundred dollars more—to my way of thinking, a very worthwhile expense if you don't have an A/V receiver/amplifier.
- And, of course, there's remote control. There's hardly a VCR on the market that doesn't have remote.

Utilizing Your Old Equipment

Finally, there's the matter of blending your existing equipment into your own system (covered in greater detail in Chapter 1, "What Goes into a Home-Theater System"). There's hardly anyone who doesn't have an old color TV or a stereo set. If you have a serviceable TV, why not hang onto it awhile longer while building a better home entertainment system around it? Later, when you have more money to spend, you can replace the TV. In the meantime, this is the most inexpensive way to get video.

If you have an existing stereo setup, check out the receiver/amplifier. You may be able to get stereo sound from either your television set or your VCR. (See Chapter 9, "Putting It All Together.") You can buy a Dolby Surround Pro-Logic decoder to go with your existing equipment (although today it probably costs more than a

new, complete A/V receiver/amplifier). Also, remember that most old units don't have remote control, which I consider an essential ingredient. This is the one area where buying new probably will be a necessity.

You might be able to utilize your current speakers for satellite purposes (left and right) and purchase only center and surround speakers. This will save money, but probably will result in unbalanced front speakers. If you can tolerate this, more power to you. I suspect, however, that eventually you will want to replace all your speakers with a matched set.

An existing VCR will work fine, provided it has four heads and stereo. If not, you'll probably want to buy new.

Building the $1,500 System

Having thus gone into the details of where to save money, what can you do to come up with a system for around $1,500?

There are many combinations of excellent equipment from which you can choose. The following was chosen simply because it was familiar to me and I know it works, sounds good, and fits the right price tag. Table 3.1 shows a minimal new home-theater system for under $1,500:

Table 3.1. Minimum Home-Theater System

Item	Model	Price
Television	Sanyo AVM-2553 25-inch	400
or	Fisher PC 3525 25-inch	430
Speaker System	Boston Acoustics	
	404v Center speaker	130
	Two HD7 satellite speakers	200
	Two HD5 surround speakers	150
Receiver/Amplifier	Technics SA-GX350 Dolby Surround Pro-Logic	300
VCR	Any brand, 4-head, hi-fi	300
	TOTAL SYSTEM COST, approximately	1,500

Figure 3.1 shows the Technics SA-GX350 Dolby Surround Pro-Logic receiver/amplifier. Figure 3.2 shows the Boston Acoustics speaker set.

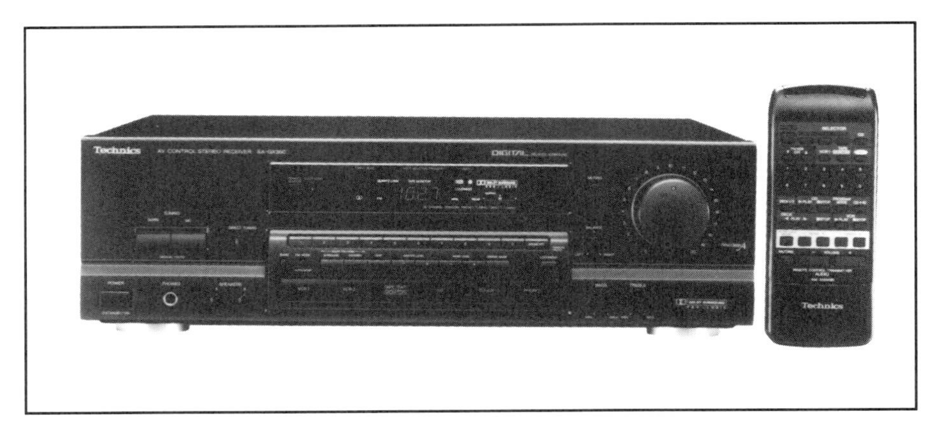

Figure 3.1. Technics SA-GX350 offers Dolby Surround Pro-Logic circuitry and equal power across the front sound state for $300.

Upgrading: The $15,000 System

Having thus established the base line for our home theater, the sky's the limit for what we might want to move up to. For now, however, let's consider a system that costs around $15,000. What would be different?

The Television Set

Both RCA (the G34168ET and the ProScan 34190) and Panasonic (the PT-50WXF5) have introduced wide-screen, 16:9 television sets for around $5,000. The RCA/ProScan model is a *direct-view* (that is, tube-type) set, 34 inches in diagonal width. The Panasonic (shown in Figure 3.3) is a rear-projection type set with a 50-inch screen. If you have the bucks, you might just want to try one of these. For the same money you can get a Mitsubishi 40-inch direct-view set—the first mass-produced 40-inch picture tube. Of course, all of these televisions include many other features. If the large screen is what you're after, however, they come close to being the ultimate.

Figure 3.2. Boston Acoustics speaker set.

For high-quality televisions, it's hard to beat Sony Trinitron XBR series TVs, which run from around $1,400 (KV-27XBR26) to $2,600 (KV-32XBR76). Hitachi's new, high-performance, rear-projection TVs are in a similar price range. Toshiba, Sharp, Panasonic and others produce excellent television sets. A Panasonic large-screen TV is shown in Figure 3.3.

Expect to pay around $5,000 for the best-quality big screen that gives great performance. You can pay less for smaller sets.

Figure 3.3. A Panasonic TV.

Better Speaker Systems

If you have the bucks to spend, you'd be crazy not to get a Home THX approved sound system, which means speakers and controller. They actually don't cost all that much more, depending on the brand and model.

For a top-of-the-line model you can spend $12,500 for a JBL Synthesis Two loudspeaker system (Figure 3.4) or $27,500 for a complete cinematic/music system. On the other hand, more reasonably priced would be an Altec Lansing Home THX system, including shielded left-, right-, and center-channel speakers, subwoofer, and surround speakers for a system price of $3,000. Boston Acoustics offers a Home THX speaker system, including left, right, and center speakers, dipole surround speakers, and two subwooferss for a total price of $2,400. Add another $600 to $800 for the controller.

Figure 3.4. The JBL Synthesis Two loudspeaker system

Many other manufacturers offer similar products. Expect to pay around $5,000 or more for a superior Home THX speaker system.

The A/V Receiver/Amplifier

As always, there are many brands and many excellent products. For example, the Onkyo Integra TX-SV909PRO with 260 watts/channel dynamic, Dolby Surround Pro-Logic, S-VHS inputs, and subwoofer output sells for $1,800. (These features are discussed in subsequent chapters.) Denon, a long time manufacturer of excellent sound equipment has units ranging from $900 to several thousand dollars. Sony, Panasonic, Harman-Kardon, McIntosh (shown in Figure 3.5), and a host of others offer excellent, top-of-the-line equipment.

For an A/V receiver/amplifier that will make your home theater sing, you can reasonably expect to pay up to $2,500.

The VCR

Sony, Panasonic, and many other manufacturers offer excellent lines of VCR products. For home use, the SLV-R1000 at $1,300 is an excellent choice.

Sony and Panasonic also offer semi-professional VCRs with *jog shuttles* (for precise control of tape positioning) in the Super VHS line. Sony offers an 8mm VCR as well. These VCRs are in the $1,000 to $2,000 price range. Expect to pay around $1,800 for a top VCR.

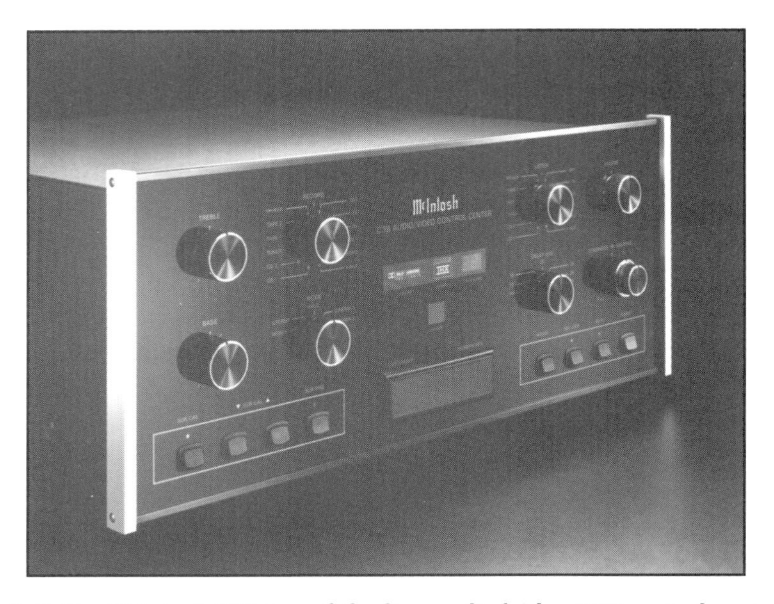

Figure 3.5. The McIntosh C39 top-of-the-line audio/video THX control center.

All That Other Equipment

Of course, we haven't included other equipment that many feel is absolutely essential to a home entertainment system. Take, for instance, the CD player.

Today you can get an adequate single CD player for around $200. On the other hand, many of us prefer to have a CD player that holds more; for example, a five-CD carousel unit. However, for a top-of-the-consumer-line model, consider the Sony 100 disc CD changer! (Figure 3.6) It holds 100 CDs so you can keep all your music stored in the changer. You don't need a separate rack. The retail price is $1,200.

As your system gets better, you will want to add a laserdisc player. Laserdisc players can be purchased for less than $500.

As noted earlier, there are also other pieces of equipment you may want, such as a CD-I player, satellite system, camcorder, and more. These are all covered in later chapters.

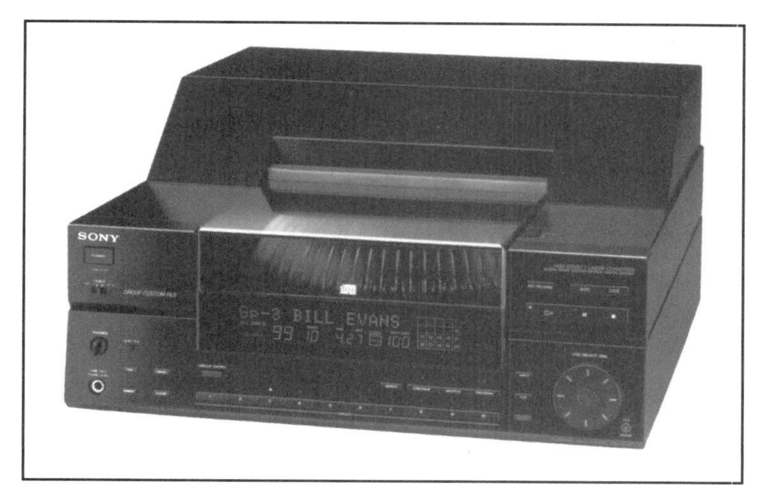

Figure 3.6. The Sony 100 Disc CD Changer.

Adding It All Up

For a $15,000 top-of-the-line system you can easily pay the following:

Big screen television	$5,000
Home THX speaker system	5,000
A/V receiver	1,500
VCR	1,000
Other equipment	2,500
	15,000

Rest assured, you can reach $15,000 very quickly without a lot of effort!

Complete Systems

One area we haven't yet covered is that of complete systems. Some manufacturers offer complete, matched home-entertainment systems for a single price.

For example, Sony's SA-VA3 (shown in Figure 3.7) is a self-contained Dolby Surround Pro-Logic sound system. The amplification, by the way, is contained in the left and right towers, which also power subwoofers located in the base of each tower. The complete system price is $950. When you add in a big-screen television for another $500, you just found yet another way to get a home system (sans VCR) for around $1,500.

Figure 3.7. A Sony SA-VA3 self-contained set.

Other manufacturers offer complete home-theater systems including TVs. The advantage here is the price, which usually is much less than buying the equipment individually. The disadvantage is that you don't get to mix and match to get exactly what you want.

CHAPTER 4

The Big Screen

What most distinguishes the movie theater experience (besides the popcorn, fantastic sound, and bad seats) from watching films at home is the big screen. When the lights darken and the curtains widen to the side, you're almost enveloped by the movie screen. (If you sit in the first five rows, you literally are craning your neck up and down and from left to right to see the whole picture. I know people who will purposely sit up front just for the dubious pleasure of getting this effect.) At the movies, much of the feeling you get of being drawn into the picture—assuming it's a good flick—comes from the screen size.

If you're a purist who believes that screen size is irrelevant and that it's the story being told that counts, try watching your favorite movie on a four-inch color LCD monitor. You'll be able to follow the plot, but you'll miss most of the cinematic effects, and probably find yourself yawning a good part of the time.

The big screen is part and parcel of the theater experience and, as a result, everyone who builds their own home theater wants a big screen. The real question thus becomes: What kind of screen should you get? (Not how big, but what kind. In Chapter 2, "Fitting the System In Without Buying a New House," we went into detail on how to determine the optimal size of screen to get.)

Before getting into types of TV sets, let's consider two universal problems with today's TV: lack of clarity and poor color. "Clarity" means how easy it is to see what's

on your TV set. It's measured by "resolution" a technical term that refers to the number of lines, either vertical or horizontal, on a TV set. "Color" means how accurately colors are reproduced. These three elements are discussed in the next sections.

Clarity

Here's a simple test you can perform to determine the clarity of your set. Turn on your color TV and wait for a commercial that involves on-screen letters. Typically, commercials show people in all sorts of indoor and outdoor scenes, but contain very little in the way of writing... except at the very end when the advertiser shows the product. Next time you're watching a commercial, try to read the words when the product appears, particularly the small type.

Keep in mind that if you already know what the product is, you'll recognize the logo regardless of whether it's clearly visible or not. However, the blurbs and lettering on the product may be unfamiliar to you and, consequently, are a real test of your TV's clarity. My guess is that you won't be able to read much of the smaller type, even if you have good glasses. The small type will seem to vibrate, and its colors, after you look at it for a while, may seem to bleed into surrounding areas.

Of course, unless you look for a lack of clarity, you may think everything is clear. We've grown so accustomed to poor-quality images that we tend to accept them as normal.

Consider: if the product is shown full screen, it's about a foot tall on a standard, 19-inch diagonal set (whose height is roughly 12 inches). That means that even the small type may be at least one-half inch in height. Now, a person with normal, 20/20 vision should be able to read type one-half inch tall at about 20 feet. So why can't you read type on TV that's half an inch tall?

"Because," you may say, "it's on TV."

So what? All that means is that you've gotten used to poor images on-screen. For a comparison, consider a high-quality computer monitor. At 20 feet a person with normal eyesight should have no problem reading the letters.

Resolution

Resolution in television sets is clarity measured in horizontal lines. This actually means counting vertical lines horizontally across the set. The classic (and still best)

method of explaining this is to imagine a picket fence, as shown in Figure 4.1. Each picket is one line of resolution. The quality of the TV set is how many of the pickets you can distinctly see on the screen before they all blur together. The more pickets, or lines, the higher the resolution of the set.

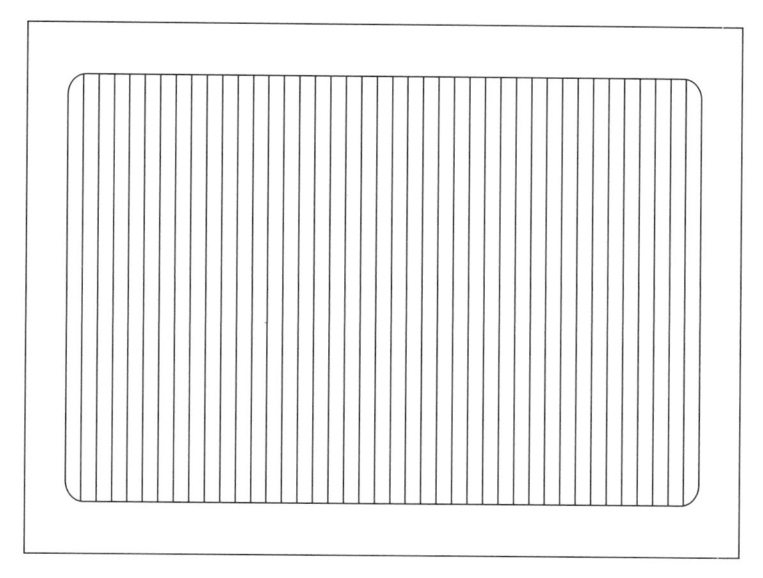

Figure 4.1. Picket fence screen diagram.

That takes us back to the NTSC, National Television System Committee, which set the standards for resolution. It's not that those people needed eyeglasses; they were intelligent and clear-seeing and they wanted the best for us. The problem was that, back in 1953, the biggest picture tube any one of them had ever seen was around six inches diagonally. Most never anticipated that television would ever have a TV set bigger than 13 inches diagonally. Consequently, the issue of better resolution never really came up.

Getting Better Resolution

One way to get a clearer picture is to increase the *vertical resolution*, or the number of times the electron tube scans the screen per second: top to bottom, zigzagging across the screen. (Tube scans occur 60 times per second in two fields of 30 scans each.) Some TV monitors scan at twice this speed, and thus are clearer to view.

Line Doublers

A number of *line doublers* are available on the market. The Faroudja LD100, for example, retails for around $2,500 and doubles the scanning lines. In the process, it actually makes the *matrix* (the lines visible on your screen) disappear, producing a smoother and somewhat clearer picture. It remains, however, a big price to pay for a relatively small increase in viewing clarity.

Horizontal Lines of Resolution

The more common method of increasing resolution, however, is to increase the number of horizontal lines: that is, get more pickets visible on-screen. This is harder to do because of the NTSC broadcast standards, which actually prohibit broadcasts of higher horizontal resolution. (This is to make sure that older sets continue to work with today's broadcasts.)

However, TV manufacturers are not restricted in the lines of resolution they can build into their TV sets. Modern, high-quality sets often are rated at 500, 700, even 1,000 lines of resolution! The problem is, of course, that you won't see that if what's broadcast is much lower, about 330 lines showing up on your screen at best and often much less.

Another way of getting increased resolution is to record a video with a Hi8 or Super-VHS (S-VHS) camcorder. These record at around 400 lines of resolution. In addition, they split the composite signal coming to the TV set into "Y" (for black and white) and "C" (for *chrominance*, or color). Some modern TV sets are set up to receive separate Y/C signals, and can display the result as true 400 lines of resolution. If you compare this signal to standard broadcast, you will be able to tell the difference instantly.

A better alternative is *high-definition TV* (HDTV), which gets to the set at a much higher resolution, 500 lines or more. However, an HDTV standard has not yet been adopted by this country (largely because of insistence that it be compatible with existing, low-definition TV sets). According to experts in the field, however, HDTV sets could be out and receiving signals within two years of adoption of a standard. (See Chapter 14, "Coming Soon to Your Home Theater," for more information on HDTV.)

If you ever get a chance to see an HDTV set in operation, take a good look. It takes only a few seconds of watching HDTV to spoil you forever on regular TV. The difference is enormous. It's like being someone with terrible vision who has been seeing

the world through a blur and suddenly acquires a good pair of glasses. Suddenly, you can clearly see half-inch text on the screen from 20 feet away—probably from 30 feet!

Megahertz versus Resolution

While some manufacturers describe their resolution in terms of horizontal lines, others do it in terms of megahertz (MHz). *Megahertz* refers to the frequency with which the electron gun (located at the back, center of the TV tube) shoots. The higher the megahertz, the better the resolution.

The rough correlation between megahertz and horizontal lines goes something like this: for every one megahertz of video signal, there are roughly 80 lines of horizontal resolution. However, our old NTSC gets into the act again by restricting the standard TV to 4.2 MHz. Thus, the best resolution you're likely to be able to get on a standard TV set is 336 lines (actually, for technical reasons, much less than that on the screen).

Color

Much of the reason for the blurriness of modern TVs goes back again to the NTSC (sometimes deridingly dubbed "Never Twice the Same Color"). In 1952, that committee established the standards for the images we watch today, including the standards of red, green, and blue that would blend to produce color. The problem was that the image on the screen is produced by activating phosphors that are placed on the back of the screen—and there weren't good phosphors for those colors. In the old days, viewers used to complain regularly that the "flesh tones were off." Actually, all the colors were off—but most people's eyes couldn't discern this.

As a result, every manufacturer used a different phosphor, and no two TVs gave exactly the same color. Further, for the color system to be compatible with black-and-white sets, it was necessary for black and white to be obtained from a combination of the RGB (red, green, blue) colors. Suffice it to say that any text on-screen is produced by blending colors. Unfortunately, due in part to phosphor bleeding, this process is not highly accurate.

Some phosphor bleeding was corrected by the addition of a black *matrix* (introduced more than a decade ago) around the scan lines of modern tube television sets. This enhances the contrast and increases the clarity. Nevertheless, the colors still are not perfectly clear even in today's sets.

Is Bigger Clearer?

Having thus "proven" that all current television is blurry with messy colors, the question becomes: Is bigger blurrier and messier? The answer is: not necessarily!

For one thing, manufacturers of smaller TV sets often do not optimize them. These usually are the least expensive sets and there's little incentive for the manufacturer to build in better resolution. Further, as the picture area is small, manufacturers often figure that the resolution really doesn't matter that much anyway.

On the other hand, manufacturers of larger-screen TVs often push the horizontal lines of resolution to the maximum.

In addition, some manufacturers such as Panasonic, claim that they do *overscanning* on some of their sets, where they double the number of scans pers second on their large-screen sets, thus increasing vertical resolution. In general, therefore, the bigger the set, the better the electronics driving it—and the better resolution achievable, up to certain limits.

If you get up close to a large tube or projection set, unlike an HDTV set, it's going to get blurry. But few of us do. We'll sit further back. And, because a bigger set gives us a more theater-like feeling, we tend to enjoy it more.

Bigger, then, though not necessarily clearer, often is better.

The Line Structure Problem

One difficulty with getting very close to any kind of a set, however, is that you will begin to see the the structure of lines that actually compose the picture. This is quite distracting, and will spoil anyone's appreciation of the picture. Therefore, sit sufficiently far back so that you see the overall picture, not the structure of which it is composed.

Rear Projection

For years, the only way you could get a very large screen was with a rear-projection TV. Mitsubishi, Hitachi, and others have long made 60-inch sets, sometimes even larger ones. (Only recently have 35-inch and now 40-inch tube sets been available.) A rear projection TV consists of an opaque screen and a television projector placed behind and below it. Typically, the projector, sitting on the bottom of the set, aims straight up at a mirror placed at about a 45-degree angle, which then redirects the light to the back of the screen. You see the picture on the other side.

For greater clarity, rear projectors don't use a single lens. Instead, they use three lenses: red, green, and blue. Of course, to get a clear picture, they must be precisely focused. With many sets this must be done each time the set is moved. Hence, with a rear-projection set, you don't want to be moving it from one location to another very often.

The big problem with rear-projection television sets, historically, is that they were very dim. In order to even see the picture you had to watch in a room with almost no lights on. This often was difficult to achieve—and when it was achieved, it often was tiring on the eyes. (See the discussion of ambient lighting in this chapter under "Getting the Right TV/Light Environment.")

Rear-Projection versus Tube-Type Sets

During 1993, advances in rear projection televisions increased their *lumens* (the light reaching the screen) by 30 percent or more. This is particularly the case for Hitachi, Sony, and Mitsubishi rear-projection sets. As a result (and in part because of a peculiar advance in tube sets, noted below), for the first time many rear-projection sets became brighter than tube sets!

This brightness phenomenon is hard to believe until you actually experience it. Today's rear-projection TVs no longer have to be viewed in a dark room.

At the same time, as rear-projection TVs became brighter, Hitachi, RCA, and others began using a special dark glass in their direct-view tube sets to increase contrast, and thereby clarity. (Hitachi calls it "ultra black.") The Hitachi set is shown in Figure 4.2. As mentioned earlier, TV contrast increased enormously when a black matrix first was used behind the screen. This put black between the lines of phosphors to help define them more clearly. Recently, it's been discovered that adding black to the glass in the front of the screen helps even more. This is done by Corning Glass, which manufactures the glass for many TV tubes in the United States.

However, because the glass is black, the picture actually is somewhat dimmer. As tube sets are very bright anyway, this loss is negligible. What this means, however, is that now rear-projection sets have become competitive with tube sets in brightness.

If you're trying to make a decision between a rear-projection set and a tube set, don't let brightness be the deciding factor anymore. Check out the sets. You'll be surprised.

Figure 4.2. Hitachi black-screen TV.

Viewing Angle

According to the experts, the ideal viewing angle is no more than 15 degrees right or left of perfectly perpendicular to the screen, as shown in Figure 4.3.

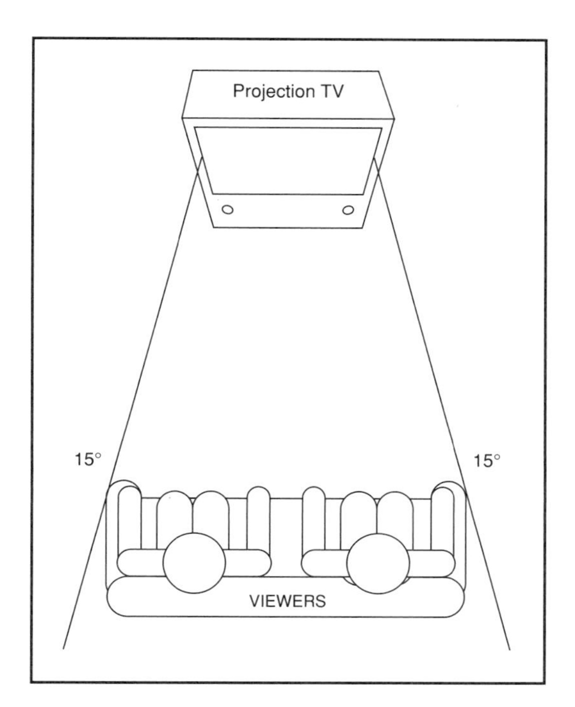

Figure 4.3. Viewing angle diagram.

The idea here is *not* that if you're more than 15 degrees off, you won't be able to see the screen. Rather, it's that you won't get the correct video effect from the film you are watching. In most movie theaters, very few seats are to the side of the screen. There are side seats; however, the edge of the screen itself is usually not more than 15 degrees off from these seats.

Modern tube sets; however, can be seen (though not well) from far more than a 15-degree angle. Indeed, in an extreme case it's possible to view them from almost a 90-degree angle to the side. Modern rear-projection TV sets, however, have a much narrower viewing angle; much past 45 degrees and the picture suddenly winks out. (It actually has to do with the linear nature of the screen, which is designed to sharpen the picture when aimed directly ahead.) This is, however, much better than older rear-projection TV sets, which had even narrower viewing angles.

In any event, be aware that rear-projection TVs generally have a much narrower viewing angle than tube sets. If you get rear projection, you'll want to keep your audience at not more than 15 degrees off center for ideal viewing. More than that and someone isn't going to enjoy the picture quite as much.

Front-Projection TVs

Front-projection TVs have been around as long as rear projection, but haven't been as popular. The size you can achieve with front projection, however, is almost unlimited. I can recall being at the Consumer Electronics Show in Chicago a few years back when one manufacturer demonstrated a front-projection set that was as large as a big movie-theater screen—more than 12 feet tall. Mitsubishi for years demonstrated a full-wall (over eight feet tall) front-projection TV—at a cost, as I recall, of close to $50,000!

As with rear-projection television sets, the problem was always low brightness. The sets were very dim and it was difficult to watch the picture.

Figure 4.4. A front-projection TV set.

Within the last year, however, this particular problem appears to have been somewhat overcome, and new front-projection TVs are extremely bright. You now can view them easily in a room with normal lighting.

Keep in mind, however, that with a front-projection TV you need a screen. You cannot simply project onto a wall and expect a good picture. In addition, you need a location for the set (as shown in Figure 4.4). Problematically, that location usually is the center of where you want to sit! To solve this problem, it may be possible

to mount the front-projection TV set from the ceiling, although this could result in some distortion of the picture. (Most movies seen on transcontinental aircraft flights utilize a front-projection TV set mounted in the ceiling.)

Comparing All Three

No one type of set—tube, rear-projection, or front-projection—is perfect for all applications or audiences. For larger screens, you'll definitely want a projection set. For smaller rooms, a tube set may be desirable. For a very large room, a front-projection unit may be your best bet.

Keep in mind that prices for all televisions have come down in recent years as technology has leaped forward. By shopping around you may be able to get the set that's ideal for your needs at a price that won't kill your pocketbook.

Getting the Right TV/Light Environment

For years people have said, "Don't watch your television set in the dark. It will hurt your eyes!"

This is the truth, for reasons we'll see shortly. Further, watching a television set (whether tube or projection) in a completely darkened room will keep you from clearly seeing the picture. So getting the right television set is, in reality, only part of the battle: the other part is creating a visual environment where the TV you get will give you the most enjoyment.

The reason light is important has to do with your total field of vision. When you sit in a room, the television picture is only a portion of what your eye sees. In fact, it may only be about 25 percent of the total image your eye is picking up. That means that 75 percent of what you're seeing *is not* TV screen.

Now, if that other 75 percent is totally black, the iris in your eye automatically opens to maximum width to allow more light in. (This is the phenomenon that happens in the dark. Your irises automatically open to their widest to let in the most light so you can see more clearly. Look at a person's pupils in low light and you'll see that they're enormous; the iris is wide open.)

When your eyes are wide open, adjusting to the darkness of the room, they actually get too much light from the television, which only occupies about a quarter of the view. This can cause a strain on your optical nerve. That's why it's bad for you to watch television in a totally dark environment with a modern bright television set.

Because too much light is coming into the eye from the TV set, the contrast and the transitions between scenes are more difficult to see: you actually don't see as well as if the room were lit.

Further, you cannot avoid the "too much light" problem at home by sitting closer to the TV set to make your irises close down. If you get that close, the picture will be blurry—and you'll be seeing the construction lines on your set.

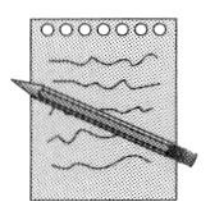

> **Note:** In movie theaters the screen is a much larger portion of the viewing area, and the light coming from the screen is much dimmer than at home. Hence, this problem really doesn't exist in the movie-theater environment. The same holds true to some extent for the older, far dimmer, rear projection large screen TVs.

Amount and Placement of Room Light

What all this leads to is that, for optimum viewing, you want there to be light in the room; more particularly, you want light that your eyes will pick up. Ideally, therefore, in a home-theater setting you will have light that shines toward the audience from behind the TV.

Of course, this ambient light must be subdued lest it interfere with the picture. Studies have shown that ambient light that is 10 to 15 percent as bright the light coming from the picture is ideal. This typically is about the amount of light coming from a 10- or 15-watt incandescent bulb.

The Color of Light

The color of the ambient light also is important. You want it to be white light (light with a high Kelvin temperature). If it's a colored light, it will distort your perception of the colors seen on your television set. (If you don't believe that the color of ambient light will distort the TVs color, try this experiment: Get a piece of pink paper, and two pieces each of yellow and red paper. (Pink is a combination of red and yellow.) Place the two pieces of yellow paper on either side of the pink paper. The pink looks more yellow. Replace the yellow paper with the red paper. The pink now looks redder.

Unfortunately, as we dim an incandescent light (which is the technique most people use to reduce the light level in a room), the color temperature changes. It becomes

lower, in the part of the scale where the reds and purples are more prominent. Thus, by lowering the light level, you may actually be distorting the colors you see.

One answer is to use a florescent light, which has a very high color temperature (white) light. A florescent light cannot easily be dimmed, however, without the use of a special and expensive ballast. However, a small florescent (5-watt) light, placed behind the TV set so that it cannot be directly seen, may actually come close to giving the desired light level as well as producing a pleasing aesthetic look. (For perfectionists, the light will not be perfectly white—it actually will have a green tinge that you may not notice, but is clearly picked up on outdoor photo film. Use of a pink gel filter—Rosco 3202— in front of the light, available at any large, quality photo supply store, should transform it into perfect white.)

Several decades ago, Sylvania attempted to address this problem by placing a florescent ring around its picture-tube sets. The brightness of the ring, however, because it was aimed right at the viewer, apparently only increased eyestrain.

There also is the matter of the color of the room itself. There's a tendency today to design rooms so that they reflect contemporary styling, which (as of this writing) often means the pinks, reds, and golds of the American Southwest. These colors also distort your perception of the color coming from your TV set.

Like it or not, the ideal color for your viewing room is gray. Gray is neutral. If you don't want to have your home theater painted a drab gray, however, consider using gray at least around your television set. You'll be surprised how much it helps.

You'll find more on this topic in Chapter 10, "Furniture and Lighting that Make the Difference."

Tuning the Color on Your Set

If you want to be absolutely sure that you're getting the right color picture on your set, you can tune it precisely. To do so, you will need to write to SMPTE, the Society of Motion Picture and Television Engineers, and ask for their "Monitor Set-Up Tape." (Specify VHS tape V2RMSV.) Play the tape in your VCR and follow on-screen directions to get the perfect on-screen picture. The price as of this writing is $32.

Write to:

SMPTE
595 W. Heartsdale Ave.
White Plains, N.Y. 10607

16:9 Aspect Ratio Screens

As noted in an earlier chapter, the latest development in TVs is the new 16:9 aspect ratio. This numerical expression is simply the ratio of the width to the height. Conventional television sets have a 4:3 aspect ratio: for every four inches in width, they are three inches in height. The new sets are nine inches in height for every 16 inches in width; that is, they have a 16:9 aspect ratio. In other words, they're much wider than they are tall.

When you first see the new 16:9s you may think they look awkward. In reality, however, they conform to 1.85:1 screen aspect of current movies and theaters. (70mm movies, however, are filmed in a different 2.2:1 aspect.) In order to get a modern movie (filmed for the big screen) onto your conventional 4:3 screen, anywhere from one-third to one-half of the sides of the picture have to be cropped! In some scenes, in which a character moves to stage left or right, the entire right or left hand portion of the film (where they aren't) is cropped. The classic impossible situation is where there are two characters, left and right on-screen, shouting at each other. On the conventional 4:3 aspect TV you would hear them from stage left and stage right, but see empty screen in stage middle!

The new 16:9 aspect screens are designed to handle modern movies. In addition, as noted earlier, they are designed in anticipation of the HDTV (high-definition television) signals that presumably will broadcast in this wider aspect. Keep in mind, however, that as of this writing an HDTV standard has not been set (except that it will include a wide screen). The electronics in current 16:9 sets, therefore, may not be able to receive future HDTV wide-screen transmissions. (Keep in mind too, though, that any HDTV standard agreed upon *must* be compatible with current NTSC television sets.) A word of caution, however. If you're in the market for one of these new 16:9 sets, be sure that you can switch it back to the old 4:3 aspect, or else you'll cut off the top and bottom of shows not intended for the wide screen.

Interestingly, the old films (those made before 1953) all were filmed by an old Academy of Motion Pictures standard that was almost square; in fact, almost exactly the old 4:3 size of conventional TVs. Your current set, then, actually is ideal for watching old movies!

The first of the new 16:9 sets were introduced by RCA (Thomson Consumer Electronics) and Panasonic (Matsushita Electronics) early in 1993. These were large-screen sets (34 inches wide) and were high priced (close to $5,000). By the end of 1993, however, much lower priced wide-screen sets are expected by a greater number of manufacturers.

If you're going to be buying a new television set as part of your home theater, you certainly will want to check out the new, wide-screen television sets. While they may be a bit avant garde for you, they seem to be the first wave of the future.

Dolby Surround and Dolby Surround Pro-Logic

The currently preferred audio system for most home theaters is Dolby Surround Pro-Logic or a simpler form of it, Dolby Surround. The reason for this is simple: home theater means coming as close to duplicating movie theater sound as possible. Dolby Stereo is used in most modern movie theaters as the preferred sound system. Therefore, it's only natural to use the home version in your house.

> **Note:** Home THX, which is similar to movie theater THX, is also available, but more expensive. (That will be covered in detail in the next chapter.)

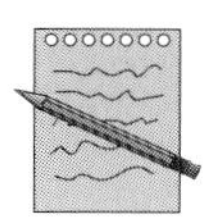

It's important to understand, however, that while we are talking about replicating a movie-theater sound, the audio in a home theater can be used for other purposes. For example, it can be used for listening to CDs, records, tapes, or other audio-only soundware. While the Dolby systems will also perform very well with audio only, they may not be absolutely necessary. You may be able to get by quite adequately

with a good, old-fashioned stereo system if all you want to do is listen. Add video, though, and the following discussion becomes vital.

In the Beginning...

Dolby Laboratories did not invent stereo or multichannel sound systems. Multichannel systems have been in use in theaters for more than 50 years. One of the first was "Fantasound," used by Walt Disney for the audio portion of the movie *Fantasia*. Others include TODD-AO, Cinemascope, and Dimension 150.

All of these systems, however, suffered from a serious drawback: in order to have the stereo sound coordinated with the movie, a special magnetic strip had to be added to the film. This was costly, and the oxides used tended to deteriorate, causing a degradation in sound quality over time.

Dolby Laboratories invented a process for recording multichannel sound optically onto movie film. This resulted in little or no degradation in the quality of the sound over time (other than the natural aging of the film). In addition, the Dolby system offered four separate channels (recorded onto two optical tracks) for superior audio performance. This was called Dolby Stereo, and almost overnight the system became the standard by which movie theater audio was judged.

In Dolby Stereo used in cinemas, there are four channels. Like standard stereo, these include left and right. In addition, however, there is also a center channel, which is used primarily for pinpointing the listener's attention to dialogue on the screen (although it also covers many important effects and music as well). The system also includes rear, or *surround*, speakers to create special effects as well as ambient sounds in the sides and back of the theater. Figure 5.1 shows Dolby Laboratories' Dolby Stereo diagram.

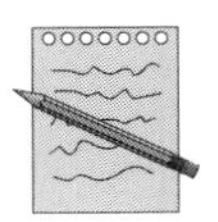

Note: Contrary to common belief, there are *not* two surround sound channels (except possibly for some 70mm cinema presentations), even though there are two (or more) surround-sound speakers. The surround portion of the sound is monaural. In some cases, such as with 70mm film, you can have "split surrounds" in theaters so equipped (very few) with releases that have the special encoding. In 70mm presentations, the two extra channels are used for bass (called "baby boom" channels). If the feature uses "split surrounds," high-frequency noise is also recorded on those channels—since the bass channels in front have a low-pass filter on them.

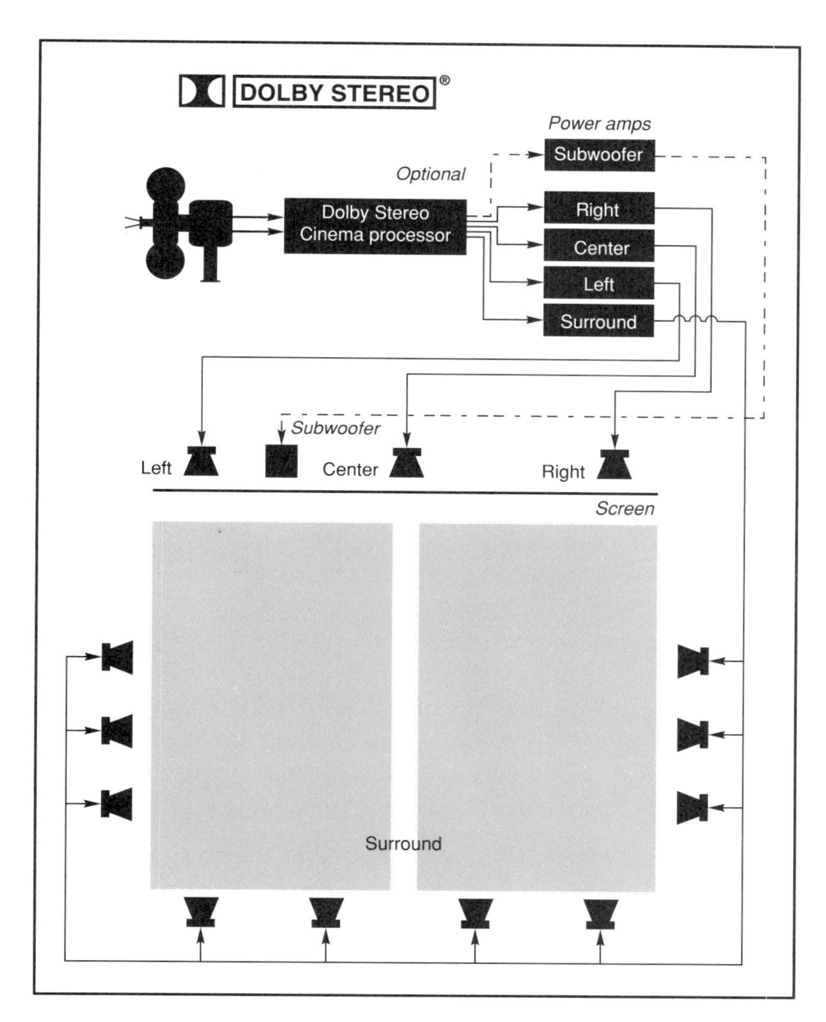

Figure 5.1. Dolby Stereo Diagram from Dolby Laboratories.

As noted, soon after its introduction in cinemas, Dolby Stereo quickly became the standard by which movie-theater sound was judged. (George Lucas's *Stars Wars* movies, released soon after, helped promote the system enormously.)

Dolby at Home

Currently, there are over 2,300 Dolby Stereo movies available on video cassettes, laserdiscs, and through television broadcast. The problem is that when these are

played on a conventional stereo system, while you can hear the soundtrack, you completely lose the surround effect.

It was inevitable that, with the success of Dolby Stereo in theaters, a home version would be introduced. That came in the form of Dolby Surround, offered in 1982. This produced sound similar to that found in a theater. Essentially, Dolby Surround added a surround effect to an otherwise conventional stereo presentation. This added the front-to-back dimension found at the movies, and gave a more complete cinema experience at home.

It's important to remember, however, that in a theater there are, as noted, four channels: left, right, center, and surround. In Dolby Surround there is no center channel. Rather, Dolby Surround uses three channels: left, right, and surround. (This is called a *passive-matrix system* because the decoding is fixed in left, right, and surround channels.) The center is a *phantom channel*, always present (even in conventional stereo systems where the sound from both left and right channels is mixed).

Dolby Surround provided a vast improvement in listening to films at home. The encoding required to replay Dolby Surround at home was already present on VHS tapes and laserdiscs of movies originally released in Dolby Stereo, and something approximating theater sound was achieved. (By the way, if you ever wonder whether a VHS tape, laserdisc, or broadcast has Dolby Surround sound, the simplest way to find out is to check on the box or look for an announcement at the beginning of the broadcast. If an announcement says it's in Dolby Stereo, you can hear it in Dolby Surround at home.)

For those who wanted exactly the same sound as the theater, the phantom channel presents a drawback. In the theater a separate, center channel is used primarily to help make dialogue clear, as well as to keep the audience focused onto the screen. The phantom channel found in conventional stereo and in Dolby Surround simply didn't work well at this. Dialogue tended to get garbled.

Thus, five years later (in 1987), Dolby Laboratories came out with Dolby Surround Pro-Logic. This created a fourth, center channel that ensures that on-screen sounds, especially dialogue, always come from the screen as in the cinema. This also allows a wider useful listening area for the viewer, as shown in Figure 5.2.

Dolby Surround Pro-Logic uses an *active-matrix decoder*. This provides additional enhancement to the directional information in order to improve the separation between all the channels. It does this by sensing and enhancing the dominant signal at any given moment and redistributing the non-dominant signals. In essence, the variable matrix decoder listens to the input for both left and right channels and electronically determines the correct location or direction of the sound. It then

sends out four channels to place the sound in the appropriate location. This directional enhancement is the reason that Dolby Surround Pro-Logic sounds more like a real movie experience than simply Dolby Surround.

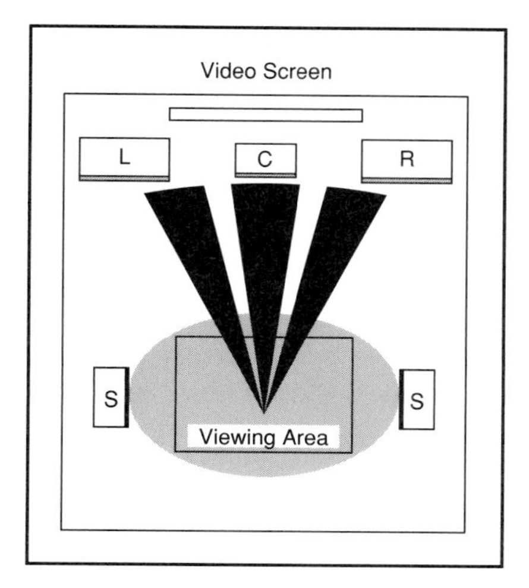

Figure 5.2. Dolby Surround Pro-Logic speaker layout and listening area. From Dolby Laboratories.

Both Dolby Surround and Dolby Surround Pro-Logic can achieve pleasing and effective results. The Pro-Logic decoder, however, provides better separation, with true center-channel output to improve the localization of on-screen sound and allow a wider listening/viewing area.

The Dolby Systems

There are other surround systems (though only licensed Dolby products carry the "Dolby Surround" or "Dolby Surround Pro-Logic" logos). The other systems work quite well. However, the trademark Dolby systems, in my opinion, work best for viewing the following:

- Movies originally recorded in Dolby Stereo and then transferred to VHS tape
- Movies on laserdisc
- Many television, cable, and satellite broadcasts, which also now use Dolby Surround encoding

Remember that Dolby Laboratories produces no consumer equipment, only the professional products used in cinemas and studios. Rather, they license their circuitry to manufacturers (such as Sony, Technics, Denon, and so on), who then produce decoders, amplifiers, receivers, and other equipment using the special circuitry. If it doesn't say "Dolby" on it, however, it isn't.

Surround Sound

What usually catches the consumer's attention is the surround speakers, the (usually small) speakers placed in the rear or sides and that carry *ambient sound*, the "other sound" that normally makes up our audio environment. Understanding how these speakers work goes a long way toward understanding the surround-sound experience. Figure 5.3 shows the typical setup for two- and four-speaker Dolby Surround Pro-Logic systems.

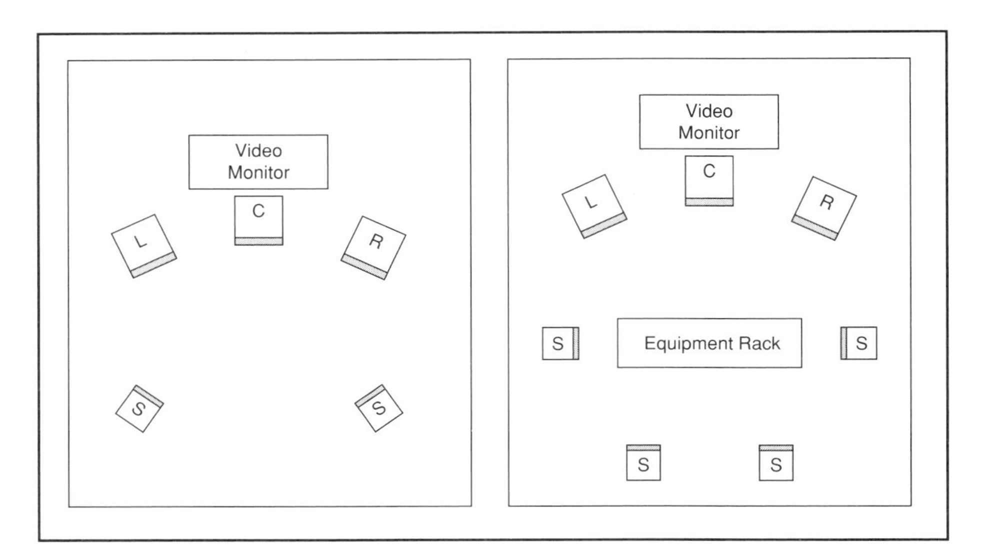

Figure 5.3. Dolby Surround Pro-Logic speaker layouts for two and four surround speakers.

The back or surround speakers carry information on a separate channel. This information is that extra sound that we normally hear in real life. For example, we may hear a bird chirping in a tree. While the bird and tree may be in front of us, some of the sound may reverberate off a house behind us. It's hearing that extra "spatial" sound that gives us the sense of realism for which surround sound is so well known.

There is, however, a problem inherent in surround sound speakers. That problem is known as *localization*. This means that, while it's important to hear the surrounding sounds, it's just as important not to be distracted by them. We want to stay focused up front, on the screen, looking at the movie. If the bird chirps behind us, however, our heads will automatically turn toward the new sound and our attention will be broken.

All Dolby Surround decoders use two techniques to avoid this problem: time delay and limited sound frequency.

Time Delay

It's a well-known psychological phenomenon that our minds associate a sound with the location where it first is heard. If it's first heard in front of us, we will continue to focus in front, even if we later hear that sound behind us.

Thus, Dolby Surround incorporates a delay in sound that goes to the rear speakers. If our bird chirps and we hear it first in front, even if we *later* hear the same chirping sound behind, we still will keep focused to the front. This is particularly important because, in any system, there's going to be some *leakage*, or sound that will be carried on all channels, front and back.

This delay is important to focusing our listening to the front. However, the length of optimum delay varies both with the listener's position and width and size of the room. Better-quality amplifiers/receivers, therefore, incorporate a *variable-delay control*. When you're looking to purchase a Dolby Surround system, check to see if it offers variable-delay control. With it, you'll be better able to "tune" the system to your room.

All Dolby Surround decoders have:

- A time delay in the surround channel. In some this is fixed at 20 microseconds, which is appropriate for most home-theater settings. Others offer a delay that is adjustable from 15 to 30 microseconds. This allows the delay to be optimized for the room.
- An input-balance control to optimize the accuracy of the decoding. Figure 5.4 shows two graphs that depict fixed and variable time-delay decoders for surround speakers.
- Separate volume controls for the adjustment of each channel, as well as a master volume control for adjusting the volume of all output channels simultaneously.

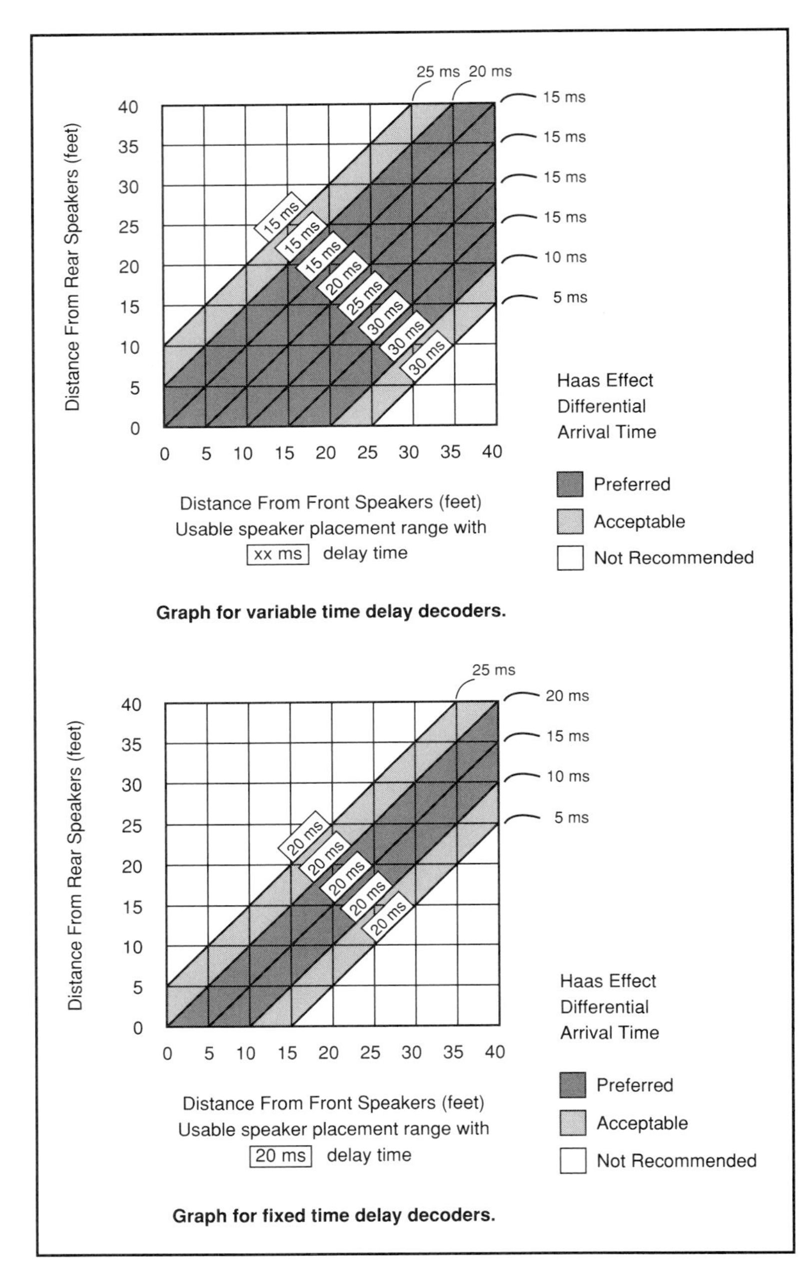

Figure 5.4. Graphs for fixed and variable time delay decoders.

Limited Sound Frequency

The second method Dolby uses to avoid localization is to restrict the frequency heard on the rear surround speakers to a maximum of 7 kHz (7,000 cycles per second) and a minimum of 100 Hz. The surround speakers thus do not handle very high or very low sounds.

You'll recall that the normal human ear hears from around 20 Hz to 20 kHz. Thus, the surround system eliminates the top 13,000 kHz, or the top octave of hearable sound.

There's a very good reason for this. Most sounds we hear are at lower frequencies, often under 4 kHz. The higher frequencies usually carry just the harmonics that help us to identify the nature of the sound. Further, higher-pitched sound is extremely directional. The 7-KHz filter removes high-frequency crosstalk, such as dialogue sibilants (the ss and sh sounds), which would be distracting. These leakages occur because of the accumulation of phase (due to azimuth misalignments or equalization mismatching errors) between the signal channels. The decoder cannot prevent these errors from going out in the surround output; therefore, they are filtered off.

In addition, the many generations of transferring from one medium to another before the signal reaches your home often introduces other errors into the system resulting in increased leakage between front and back channels. Because most of these errors occur at higher frequencies, eliminating those frequencies also helps to avoid localization problems.

Additionally, very low sounds (those under 100 Hz) are nondirectional. Therefore, these also are eliminated from the surround channel, to be carried instead by the larger front speakers.

This is important to remember when purchasing surround sound speakers. Because it's not necessary to cover deep bass, a good, "two-way" bookshelf speaker makes an excellent surround choice. You don't need an expensive surround-sound *tweeter* to carry high notes or a *woofer* to carry low ones. What you do need is an excellent mid-range speaker. These often are available for a fraction of the cost of a full-range speaker system. They are also considerably smaller.

See Chapter 7, "Speaking of Speakers," for more on this topic.

Subwoofers

As noted in Chapter 1, many Dolby Surround Pro-Logic systems include a separate output (not a real channel, but a blend of all channels) for a *subwoofer*, a special speaker for sounds under 100 Hz. This is another feature that you may want to look for, as it simplifies the connecting of powered subwoofers.

A subwoofer channel in your receiver/amplifier means that you can be assured of getting the deep, low sounds that a conventional speaker might not be able to handle. Further, because these sounds can be carried by the separate subwoofer, you can get smaller left- and right-satellite speakers. You don't need to have separate subwoofer elements in those speakers.

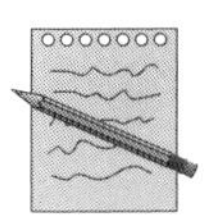

Note: Many manufacturers of A/V receivers provide pre-amplification only for the subwoofer. You may need to use a self-powered subwoofer. The same applies to the surround speakers (you may need self-powered surround speakers).

Volume Controls

If you have three or four channels with four or five speakers, you're going to need to do some tuning along the way. Keep in mind that once you've got it tuned, you don't want your system to go out of tune every time you change the volume. Therefore, it's important when buying a receiver/amplifier that you look for one that includes a *master volume control*. (All Dolby Surround systems have this.) This means that you have separate controls for each channel, and then one master control that raises or lowers the sound for *all* channels. You use the separate controls to tune the system and the master control to raise or lower the volume.

Tuning Your System

Tuning your system involves calibrating the sound levels on all the speakers so that one speaker is not blasting while another is too soft. In the traditional stereo system, there's only the "balance control" that adjusts the relative volume between left and right speakers. Here, we're doing the same thing but adjusting for center and surround as well.

The actual tuning is fairly simple and is done according to the instructions that come with your A/V receiver/amplifier. Further, in Dolby Surround Pro-Logic there's a built-in test signal (called *pink noise*—a kind of hiss on all channels) that aids enormously in the calibrations.

When tuning your system, however, some general guidelines to follow are these:

- Is the system able to play *loud enough* for you?
- Does the frequency range go *low enough* for you?
- Are the speakers sufficiently *well-matched tonally* so that you're happy with the results?

Have faith in your own perceptions. Remember that setting up the system means getting it to sound good for *your* ear, not matching it to some hypothetical ideal. The surround experience should enhance your listening experience—which, after all, is the whole point of having a home theater.

State of the Art

When you use Dolby Surround or Dolby Surround Pro-Logic, you are getting state of-the-art theater sound in your home. After all, isn't that what you're after?

6 CHAPTER

Home THX: Maximizing the Audio

If your goal is merely clear sound for playing back CDs, records, or tapes, then any conventional stereo system probably will do. Two-channel systems (with phantom images between the two) usually are quite adequate for handling most recorded audio.

On the other hand, if your goal is to achieve movie-theater audio at home, standard stereo just won't do at all. As we saw in the last chapter, you're going to need to move up to Dolby Surround at the least, and probably Dolby Surround Pro-Logic.

However, if you're the sort who wants a special listening experience, defined as bringing the absolute best movie theater sound into your house, you'll want a Home THX Audio System. This is the home version of the Lucasfilm THX theater system. Many consider it to be the current state-of-the-art when it comes to reproducing theater sound.

How It All Started

The first *Star Wars* film was introduced in 1976 and made use of Dolby Stereo encoding for maximum sound impact. However, George Lucas found that he was not satisfied with the sound in many theaters. He felt the dialogue tended to get lost amidst the dramatic music. Further, he felt that the sounds of the star fighters, jungle, and other elements were not convincingly carried by the surround speakers. So he set his corporate technical director, Tomlinson Holman, to improving it. The result was THX, a sound system that did justice to the *Star Wars* films.

In order to utilize the THX system, a movie theater cannot simply install a new audio system. Rather, it must purchase special speakers approved by THX and a special sound processor that includes Dolby Stereo decoding (a new Ultra Stereo Decoder has just come on line), and the entire system must be installed and re-checked periodically by THX engineers. Further, the theater must meet stringent acoustical requirements covering background noise, room echoes and reverberation, and noise isolation and the screen size and viewing angles must be appropriate. It's a costly investment and theaters that have THX sound, understandably, boast of it in their advertisements.

A Home THX audio system was introduced in 1990 by Lucasfilm for those who want the theater experience in their own homes. It's similar to the system used in the cinemas and utilizes Dolby Surround Pro-Logic. You can work your way up to true THX by buying bits and pieces of a system. However, in order to have a true THX system, all of the appropriate components need to be THX. (You can pair up THX and non-THX components and it should still sound okay; just not as good as the full THX system.) In order to get complete Home THX sound, you must buy a system that usually includes a special controller plus front and surround speakers.

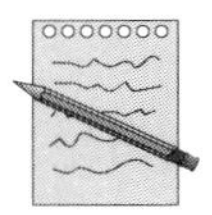

Note: Home THX does not compete with Dolby Surround Pro-Logic; rather, it is an enhancement of it. Nor does Home THX build equipment; like Dolby Labs, it only licenses the use of its processes to manufacturers.

Television Programs in THX

In order to fully utilize a Home THX *system*, the program must be recorded in Dolby Surround. While many films are recorded in this manner, recently a number of television shows have been produced and broadcast also using Dolby Stereo encoding. These include *Arsenio, Northern Exposure, The Super Bowl, World Series,* other sporting events, and even *The Simpsons*. Any show that has been so recorded will include the Dolby Stereo logo in the credits and may say at the beginning, "In Stereo, Where Available." Any of these broadcast shows can be played in Home THX, providing your local cable company transmits in stereo.

Companies Offering Home THX

As of this writing, thousands of Home THX systems have been sold, yet the system is still in its infancy. And it remains pricey. The first true Home THX-authorized system was licensed to Technics in 1990. Shortly afterward, Lexicon began making a decoder and Snell Acoustics began making loudspeakers. In 1992, Technics brought out its second-generation system, which included front speakers, surround speakers, a multichannel power amplifier, and a THX controller. The suggested retail price was $8,500. According to Lucasfilm, the licensing of other manufacturers is proceeding at a fast clip.

Companies that have been licensed to produce Home THX products as of this writing include:

Altec Lansing Consumer Products
Audio Design Associates
B&W Loudspeakers Ltd.
Boston Acoustics
Bryston Ltd.
Carver Corporation
Celestion
Fosgate Audionics
Frox, Inc.
Hafler
Harman/Kardon
JBL Consumer Products
KEF Audio (UK) Ltd.
Kenwood U.S.A.
Kinergetics Research
Lexicon Inc.
Marantz USA
McIntosh Laboratory Inc.
Miller & Kreisel Sound Corp,
Monster Cable
NAD Electronics, Ltd.
Parasound Products, Inc.
Rane Corporation (equalizer)
Snell Acoustics, Inc.
Soundstream Technologies
Technics Home Audio Group
Triad Speakers, Inc.

Uni-Screen, Inc. (Home THX screen)
Velodyne Acoustics

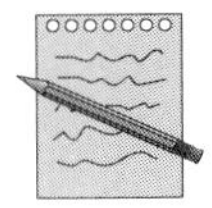

Note: Not all of these companies make full Home THX systems. Most make only speakers, controllers, amplifiers, or other elements of the system.

How It Works

Home THX was not designed as a "listen only" system, although it obviously will produce excellent audio for any reason. Rather, it was designed as the ideal accompaniment to video; specifically, films that were recorded in Dolby Surround. Thus, the goals for the system are A/V related. Home THX provides:

- Clear reproduction from the softest to the loudest sounds.
- Sound balancing across the full range of audible frequencies (20 to 20,000 Hz). It actually adds an octave above and below most other systems.
- Sharp, crisp dialogue with exceptional imaging (having the dialogue come from the location of the character speaking on-screen).
- Extremely wide dynamic range. Very realistic sound that lets you more thoroughly experience what you're seeing on-screen.

Figure 6.1 shows a diagram of a Home THX audio system. Each element is discussed in the following sections.

Speaker Setup

A Home THX system requires a minimum of six speakers: left, right, center, subwoofer, and at least two surround. (Additional surround speakers are an option.)

The speakers must meet strict standards for directivity and general performance. In addition, they must be positioned correctly in the room. You cannot substitute your existing non-Home THX speakers for those in the system and have it work perfectly. You can, however, make the substitution, and still get excellent sound.

The front THX speakers are designed to focus the sound directly at the listener. Both vertical and horizontal directivity are strictly controlled. *Vertical directivity* is

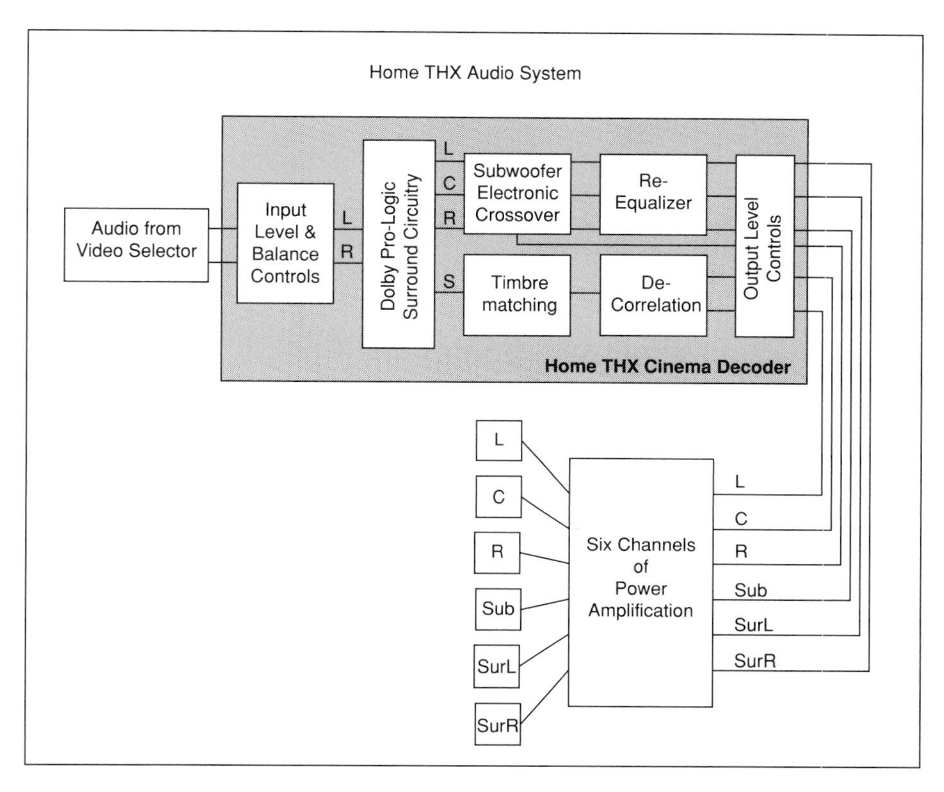

Figure 6.1. Home THX Decoding.

shaped to reduce reflections off floor and ceiling. *Horizontal directivity* is kept wide to allow a listener in other than a center position in the room to hear the sound.

The front three speakers (left/center/right) have a range of 80 Hz to 20 KHz. They're specifically designed to have a balanced octave-to-octave sound. Low frequency sound under 80 Hz (it's generally under 100 Hz for other systems) is handled exclusively by a subwoofer which has a range of 20 Hz to 80 Hz.

The surround speakers are specially designed to direct sound in two directions. They are placed at the side of the room (as opposed to the back in traditional surround systems) and direct sound to both front and rear, with a *null* (or no-sound area) directed toward the listener. As a result, those listening perceive a wide and enveloping sound field. Figure 6.2 shows a graphical representation of the sound dispersal from surround speakers.

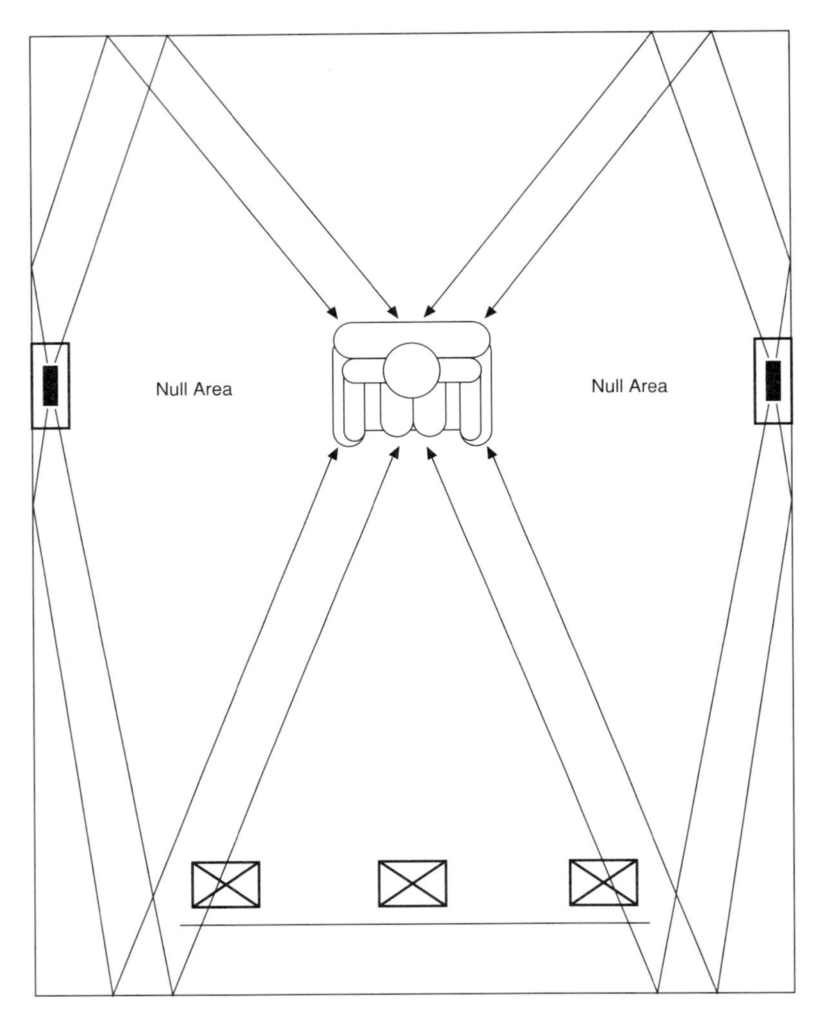

Figure 6.2. How sound from surround speakers is heard.

The Heart of the System

Home THX utilizes a controller that decodes the audio track signals using Dolby Surround Pro-Logic decoding. It then enhances the sound before sending it out through the different channels. Three separate enhancement techniques are used: re-equalization, decorrelation, and timbre matching. These are discussed in the next sections.

Home Theaters Customized to Fit Your Lifestyle

Plate 1.
Theater in a cabinet using RCA equipment.
Photo courtesy of Thomson Consumer Electronics.

Plate 2. left, top
In-wall, recessed home theater installation. Photo courtesy of CEDIA/Behrens Audio-Video, Jacksonville, Florida.

Plate 3. left, bottom
Wall installation in smaller room. Photo courtesy of CEDIA/V. Frederick International, Palm Desert, California.

Plate 4. above
Installation featuring warm wood tones. Photo courtesy of CEDIA/Behrens Audio-Video, Jacksonville, Florida.

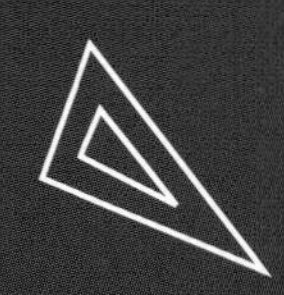

Plate 5. Versatile panel installation. Photo courtesy of CEDIA/Vaudio Automation, Valparaiso, Indiana.

Plate 6. Diagonal design for more room in a tight setting. Photo courtesy of CEDIA/ Installation: Audio/Video Entertainment, Laguna Niguel, California. Photo: Robert Freed.

Plate 7.
Flush, wall-mounted system. RCA P52152ST TV shown. Photo courtesy of Thomson Consumer Electronics.

Plate 8.
Using conventional cabinets for a classy look. Photo courtesy of CEDIA/ Intergrated Media Design, New York, New York.

Plate 9. right, top
Large screen for total home theater effect. Photo courtesy of CEDIA/Music Room, Redmond, Washington.

Plate 10.
Installation showing neutral white background for better viewing. Photo courtesy of CEDIA/ Custom Installation Services, S. Dennis, Massachusetts.

Plate 11.
Home theater close up using Mitsubishi rear projection TV. Mitsubishi offers two 60-inch Slim Line Projection TVs with a suggested retail of $3,899 and $4,399. Photo courtesy of Mitsubishi.

Plate 12.
RCA large screen TV P52152ST with a suggested retail price of $2,899. Photo courtesy of Thomson Consumer Electronics.

Plates 13 and 14. left, top and bottom
Sharp's direct view XV-120ZU SharpVision projector gives big screen performance for $495.

Plate 15.
Thomasville home theater furniture with Philips electronics. Thomasville is one of the first furniture companies to introduce home theater furniture. Photo courtesy of Thomasville.

Plate 16. below
Small room setup using rear projection TV. Mitsubishi Slim Line 50-inch Projection TVs range in retail price from $2,999 to $3,499. Photo courtesy of Mitsubishi.

Plate 17. right, top
Modular furniture available to build a home theater center. Photo courtesy of Bush Industries.

Plate 18. right, bottom
Bose Acoustimass powered speaker system places the high and mid-frequency drivers in tiny, unobtrusive enclosures, "Bose cubes." The Lifestyle 5 music system has a suggested retail of $1,449. Photo courtesy of Bose Corporation.

Plate 19.
Sound system speakers suspended from ceiling provide unobtrusive home theater. Photo courtesy of Bose Corporation.

Sound Systems Customized to fit Your Lifestyle

Plate 20. Bose Lifestyle music system offers powered speakers that are small enough to be almost invisible. The top-of-the-line Lifestyle 10 system has a suggested retail of $1,849. Photo courtesy of Bose Corporation.

Plate 21. Bose Lifestyle speaker system featuring one cube per channel instead of a dual cube array. The entry level Lifestyle 3 music system has a suggested retail of $1,099. Photo courtesy of Bose Corporation.

Plate 22.
Tiny modular unit featuring dual cubes is small enough to fit almost anywhere yet produces a powerful sound. Photo courtesy of Bose Corporation.

Plate 23.
Tiny speakers in white suit almost any decor. Photo courtesy of Bose Corporation.

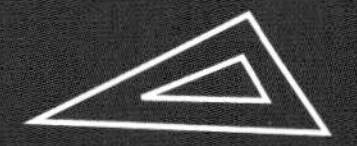

Re-Equalization

In large movie theaters, equipment standards call for high frequency sounds to be *attenuated*, or tapered off, before they reach the audience. To compensate for this, the theater sound track increases the high frequency sounds, which then appear normal to theater listeners.

Played at home, however, a film in Dolby Surround may sound hissy, or *brilliant*. There will be too many high-frequency sounds not absorbed by the small home-theater room environment. In a special process, the Home THX controller compensates and creates the original, flat-response characteristics of the recording. This happens prior to amplification.

Decorrelation

In movie theaters there are numerous surround speakers—perhaps dozens, depending on the size of the theater. The large number of surround speakers and their dispersion throughout the theater help to diffuse the sound so the listeners feel enveloped in sound. At home, however, there usually are only two surround speakers, and it's easier to get localized sounds from them.

To overcome this problem, Home THX takes the monaural channel for the surround speakers and breaks it into two outputs: one for the left and one for the right. These outputs are "uncorrelated" (not complimentary to each other), so it creates a sensation of diffusion and envelopment.

Timbre Matching

Understanding this concept is more difficult until you realize that the way we differentiate sound from different sources (say, a cello in its high register and a violin in its low register, both playing the same note) is through the *timbre*, or harmonics.

Furthermore, because of the configuration of our ears, we tend to perceive sounds differently when they come from the front (the sound bouncing off our heads and ear sides) and from the side (where the sound more or less goes straight down the ear canal). Speakers, in turn, because they're in different positions relative to our ears, are perceived as having their own characteristic timbre. Sound moving from one speaker to another appears to change in timbre, and the effect is noticeable by the audience.

Lucasfilm uses the example of an airplane moving overhead. As the engine noise passes from the front speakers to the surround speakers, the audience perceives a different timbre in the speakers.

To overcome this problem, Home THX filters the signals as they move from front to back and reduces the perception of a change in timbre.

Amplifiers

In addition to the controller that handles re-equalization, decorrelation, and timbre matching, Home THX sets strict standards for power-amplifier construction, ensuring that Home THX power amplifiers provide high-level, low-distortion outputs. Further, the amplifiers must provide stable outputs under a wide operating temperature range and variety of control combinations. The amplification usually is a minimum of 100 watts per channel for normal rooms, and more for larger settings.

The Ultimate in Home Theater

The goal of the Home THX System is to bring you the finest audio for video. If you opt for the system, you won't be disappointed.

However, I have heard a number of people express concern about buying a very expensive audio system that's only for use *with video*. Most of us want a universal system that will handle all of our audio needs. Will Home THX measure up in this way?

The answer, I have found is a resounding *yes!*

Regardless of your musical source, Home THX does several things quite well. It makes any dialogue (including singing) quite clear. You won't have the problem found in some systems (even expensive ones), where the music is delightful but you can't understand a word of the song.

Second, the bass sounds are superior in Home THX. I mean this in three ways:

- First, the cross-over at 80 Hz seems to work well with little or no leakage. Other systems that have a higher cross-over or leakage send some sound, including low voice, to the subwoofer and amplified. This can be extremely disturbing to listen to.
- Also, the Home THX system provides enormous bass response. You can feel the vibration as those ubiquitous airplane engines pass overhead. This has the added benefit that when you're playing orchestral, operatic, or even rock music, you'll get the full feeling of sound depth that the music generates.

- Finally, the dipolar surround speakers add nondirectional ambient sound to any audio presentation. If you want to feel like you're there at a rock concert, listen to the sound on Home THX.

A Home THX sound system may be completely out of range for you at this time. However, if you're aiming at building an ultimate home theater, keep it in mind for the future.

CHAPTER 7

Speaking of Speakers

Many people feel that the most important part of their home entertainment system is the television. After all, they note, it's what you see that counts.

I would dispute that. Try this experiment. Turn on your television set, but leave the volume turned off. Now try watching a show, *any* show. I'd wager that within five minutes you'll be bored to tears and within ten minutes you'll turn it off.

Without sound, video becomes unconnected heads, bodies, and scenes that just don't make any sense, particularly on small screens. (With large screens, you can actually follow some of the dialogue by watching the mouths of the actors.) For most people, it's the audio that ties everything together. Audio is just as important as video, and maybe even more important. After all, before television there was radio, and it prospered for decades without any visual reference.

If audio is important, few would argue that the heart of any audio system is the speaker. It's what generates the sound. (This is not to say that the amplifiers and receivers aren't also important, only that the speakers are vital.) If there's any one area of your home entertainment system where you don't want to scrimp, it should be your speakers. Simply put, buy the best speakers you can afford.

The Number and Type of Speakers

Most people already have speakers of one sort or another, and when deciding to put together a home entertainment system they anticipate incorporating their existing speakers into the system. Often this can be done quite effectively.

How Many Speakers?

A conventional setup requires only two speakers: left and right. Ideally, they should be matched so that the sound is similar and one doesn't dominate over the other (although the sound level can be compensated somewhat by the balance control at the amplifier). Perhaps you already have two such speakers. If so, you may be well along your way to a home-theater setup.

Conventional stereo uses two channels: left and right (although technically the word "stereo" does not mean just "two"—in cinemas it means "multichannel"). The left and right channels create a *phantom channel* in the center, which is a mixture of both left and right where phantom sound images can be heard. To listen to such a system properly, you should be seated at the point where the axis of both speakers cross: in the center. This setup works well for audio-only music. However, as noted elsewhere in this book, add video and you have other concerns—such as getting the dialogue to come from roughly the same place the characters on-screen are speaking (called *stereo imaging*), and getting it to sound right.

Surround speakers are essential to any home-theater setup. With Dolby Surround, this adds an additional monaural channel that requires at least two speakers for dispersion, placed at the sides usually near the back of the room, often up high. They provide ambient or fill-in sound to create a sense of realism, and also carry extra information. Thus, for a minimal home theater you'll need four speakers: left- and right-front and two surround in the back.

Today, however, most enthusiasts are moving up to Dolby Surround Pro-Logic, which provides an additional center channel for better stereo imaging. This means that you'll need another front-center speaker. Therefore, a typical home-theater setup today actually involves five speakers.

For a more advanced system, add a subwoofer for a total of six speakers. This provides better deep bass sounds and allows the use of smaller left- and right-satellite speakers. An Altec Lansing subwoofer is shown in Figure 7.1. Subwoofers typically carry only sounds under 100 Hz.

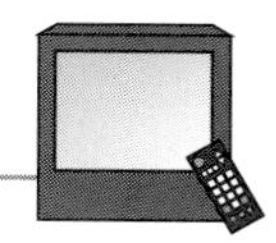

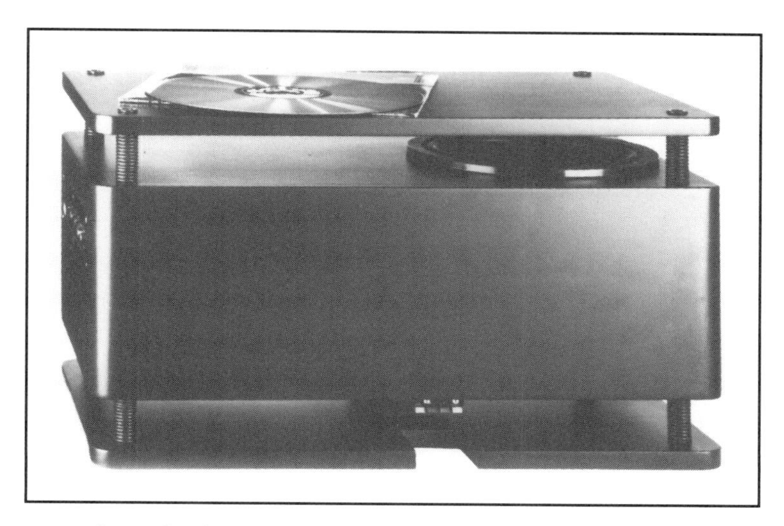

Figure 7.1. Subwoofer for those deep bass sounds from Altec Lansing. Lists at $1,200.

Modern Home-Theater Speaker Requirements

Left speaker
Right speaker
Center speaker
Left Surround speaker
Right Surround speaker
Subwoofer

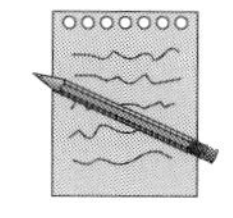

Note: Although there are six speakers, there still are only four channels. There are separate channels for left, center, right, and surround. All surround speakers are basically monaural. The subwoofer gets all the information going to the front speakers. However, because the cutoff is at 100Hz, it handles only very low, nondirectional sound.

Some enthusiasts prefer five speakers in front (two left, two right, and a center) and three or more surround speakers in back. There really is no limit to the number of speakers you can add to your system, although the number of channels remains the same.

Satellite and Center Speakers

It's important to understand that when we say "speaker," often what we're talking about is a "speaker system". If you take the cloth cover off the front of a speaker cabinet (the cloth fronts usually pop off), you'll see that the "speaker" typically includes a larger woofer element, as well as a tweeter element and perhaps a mid-range element. No one individual speaker can cover the entire range of audible sound. Also, there may be a combination of different types of speaker elements, from dynamic-cone woofers to planar mid-ranges and tweeters. Speaker types are described at the end of this chapter.

The type of speaker system you get should be determined by its use. The speaker systems with the widest range will be those in the front: left, right, and center. Assuming you *do not* use a subwoofer, these should carry the full audible range (from 20 Hz to 20,000 KHz). They should be the highest-fidelity transducers you can find.

As noted, if you do use a subwoofer to carry the very low sounds, your front speaker systems don't need to have as large a woofer. Thus, besides improving the sound, adding a subwoofer can help with installation problems by reducing the size of the satellite speaker systems. Figure 7.2 shows two front speaker systems and a subwoofer.

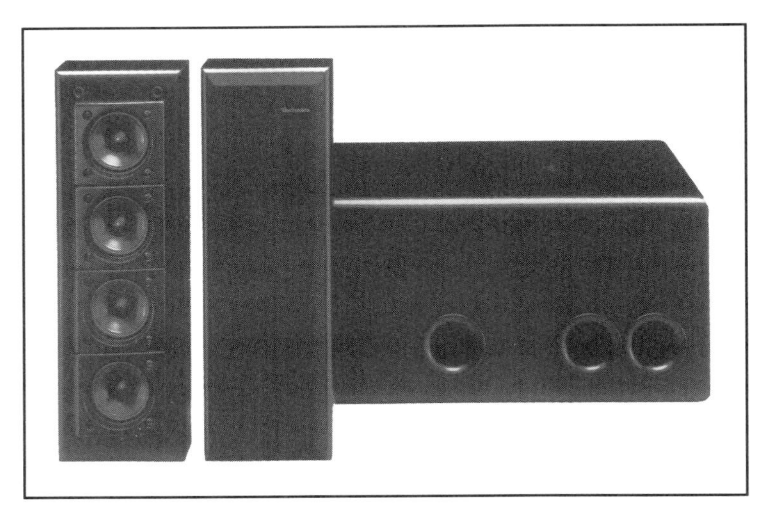

Figure 7.2. Technics Model SB-FW40 package, including two front speakers plus a subwoofer.

It's important not to think of the center speaker as a stepchild. While its main function is to help carry dialogue, the human voice drops down into the very low frequencies and its harmonics go up into the very high frequencies. To get the whole range, you need a very high quality center speaker system as well.

Center-Speaker Cautions

Caution: Do not rely on your television's built-in speakers for your center speaker channel. Often these cannot be directed by the center channel. When they can, their placement (typically at the sides of the set) does not lend itself to good center speaker positioning. Further, they often are of lesser quality than your left and right front (satellite) speakers.

The exception here are some large-screen TVs, both tube and rear projection, which have high quality speakers built-in directly below the screen. These may work as center-channel speakers.

Your center channel speaker must go where the picture is. It should be placed either directly above or below the screen. One manufacturer, Uni-Screen, produces a front projection screen (AS 1000) that is *acoustically transparent*, so you can place the speaker behind it. While most manufacturers recommend placement above the screen, my feeling is that this is just for convenience. It's easier to put the speaker on top than to build a cabinet to hold it below. However, I believe the center speaker ideally should be placed below the screen for better stereo imaging.

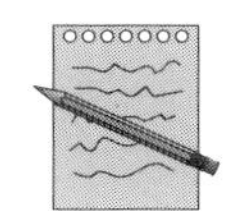

Note: Placing an older, unshielded speaker on top of a direct-view set can interfere with the picture, producing discoloration and image distortion. If you plan to place your center speaker close to your screen, be sure the screen is adequately shielded from stray magnetic fields.

Most modern speakers, however, are shielded. Figure 7.3 shows a set of shielded speakers from Design Acoustics.

Surround Speakers

Surround speakers, which go at the sides and/or the back of the room (or directly to each side, if they're dipolar—aiming to front and back in Home THX units) need not be of as demanding a quality as the front speaker. The reason is that for Dolby Surround and Dolby Surround Pro-Logic surround speakers do not carry very high or very low sounds due to a cut-off above 7 KHz and below around 80 Hz.

Thus, what you need for surround are small, mid-range speakers. If you're spending $300 to $500 for your front speakers, you can typically get a surround speaker for around $100.

Be aware, however, that the entire field of home theater is relatively new, and many speaker manufacturers do not yet offer specific surround loudspeakers. Instead, when you ask them about surround, they may simply take one of their smaller satellite speakers (designed to go up front) and say, "This will work fine."

Money-Saving Tip: The problem is that it may be too good a speaker for surround. Remember, you don't need the very high or very low tones in surround. If you pay for a speaker capable of carrying these, chances are you're paying too much (and getting a speaker that's too big in size.)

On the other hand, if you're setting up a Home THX system, you'll need dipolar surround speakers that direct sound to front and back leaving a null area at the side. These tend to be precisely built and, as a consequence, more expensive (although, again, they do not need to be full range speakers).

Tip: Why not use your old bookshelf speakers for surround when you update your system? Chances are they're quite adequate in the mid-range, may not be too big, and certainly the cost is right.

Figure 7.3. Design Acoustics PS-24 shielded speakers can be placed closer to your television set without causing interference with the picture.

Auditioning Speakers

If you were going to hire a lead singer for a band, you certainly would audition him or her. Similarly, if you're going to buy speakers, you should audition them as well. After all, it's not how they look that counts (although a good appearance is important for the home-theater effect). What you're most after is great sound. Therefore, you should listen carefully to speakers before you buy them.

A word about paying too much attention to specifications and speaker design is in order here. Over the last 10 years there has been enormous advancement made in

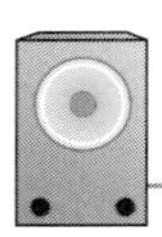

the design of speakers. Today, speakers a fraction of the size of their predecessors can create marvelous, deep sound. However, unless you have a specific design need for small speakers, you should pick your speakers more on the basis of the quality of the sound they make and their cost than on their size.

Further, manufacturers tend to overwhelm buyers with information on the technical aspects of their speakers. "Flat response from 20 Hz to 20,000 Hz" and "Seamless crossovers" are typical of the hype used to promote speakers. Be aware, however, that it really doesn't make any difference what the specifications are for a speaker. It's how it sounds to you that counts.

Finally, when it comes to your ear, don't sell yourself short. Many people feel they don't have the knowledge or experience to tell good sound from bad. Nonsense! The human ear is ten times better at discriminating a wide range of sounds than most scientific instruments. Besides, if the sound isn't pleasing to you, what difference does it make if someone else or some piece of test equipment tells you it's wonderful? The ultimate test, the *only* test, is the personal one that you make.

When Possible, Do It in Your Own Room

The sounds speakers produce are to a large degree dependent on where you hear them. Some speakers do very well in a big, wide, open area. Others do better in smaller, confined spaces.

Try to listen to the speaker in an area roughly the same shape and size as your room. But be careful. Just because a speaker sounds great in a wide, open area doesn't necessarily mean that it will sound great in your home-theater room with the sound bouncing off walls. Sometimes a retailer will allow reasonable return or exchange privileges on speakers for just this purpose. Bring the speakers home and try them out there. You could have a big surprise.

Listen to Speakers Individually

Listen to each speaker individually. That means that you pull the plug on the other speakers while you listen to the one you're auditioning.

For the satellite (left/right) speakers, put on a musical piece, preferably something that involves violins as well as basses or guitars (use a CD if possible to get more precise sounds). What you should be listening for is *balance*. All of the tones should sound evenly powerful and well-defined. The high notes of the violins should come in clear as a bell. If you're not using a subwoofers, the deep tones of the guitar or bass should likewise be clear and strong.

For the center speaker, in addition, put on vocal songs as well as spoken words. Poetry readings are excellent for this. Listen carefully to both male and female voices and see if they sound natural. Watch out for "tinniness" in the voices or too much growling, such as you sometimes get from radio announcers when the station isn't tuned in properly.

For the surround speaker, try an instrumental piece with mid-range sounds. Here, because of the nature of the speaker, you won't get the deep basses or the very high notes. But listen closely to see that the in-between sounds are well defined. A mushy surround speaker can destroy the quality of an otherwise excellent audio system.

Try It at Different Volume Levels

For classical music, you may prefer lower sound levels. Rock and roll may be higher. You may think that for home theater the typical volume is likely to be in between.

Not so. If you're going to be listening to movies recorded in Dolby Stereo and played back in Dolby Surround Pro-Logic, you're going to have sounds that are extremely loud, well over 80 or 90 decibels and perhaps as loud as 100. These won't be sustained, of course, but your speakers should be capable of handling them nevertheless.

Therefore, crank up the amps (on the front speakers, not the surround speakers, which usually require lower volume). Make sure there isn't any breakup of either music or voices at higher volume. If there is, try it with a different amplifier to be sure your amplifier isn't *clipping* (peaking out too soon at higher output, which causes distortion of the sound).

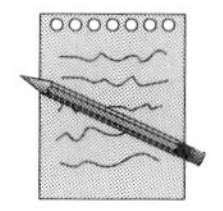

Note: Don't be confused by speakers that are rated in watts. The number of watts that goes into a speaker refers to the electrical energy required to operate the speaker. Generally speaking, the higher the wattage, the greater the sound the speaker will produce. However, the actual sound a speaker produces is measured in decibels or, more precisely, *sound pressure level* (SPL). While you want to be sure that your speakers can handle the wattage your amplifier produces (otherwise you'll blow the speaker), you want to forget all that and listen for sound levels (or use a decibel meter) when checking the loudness produced. Most front speakers today are rated at 100 watts or more. Finally, don't be confused into thinking that higher wattage means higher quality. It doesn't!

Check the Directionality of the Speakers

Contrary to popular opinion, speakers vary in their ability to project sound equally in all directions. If you're using a Home THX system, you'll find that the speakers have been intentionally built to avoid vertical dispersion while maintaining wide horizontal dispersion of sound. In other words, they'll sound good to the right and left, but not quite so good above and below normal listener height.

This is the ideal sound dispersion for a speaker—and, even if you aren't setting up a Home THX system, you should aim for this goal. Try listening to the speaker at various angles off the central axis. Ideally, you should use *pink sound* (a generated constant signal). In its absence, try to have a single tone playing or use the static hiss that you get between channels on FM.

With only one speaker on monaural and the pink sound (or its substitute) going, stand in the center. Walk to the left at least 35 degrees, then to the right 35 degrees. The high frequency of the tone (or hiss) may diminish slightly—remember, high frequencies are extremely directional—but the mid-range and lower ranges should remain constant. If you notice any sort of dramatic change, it indicates that the speaker does not have equal horizontal dispersion of sound (off-axis) and that someone sitting to the side of the room will not hear as well as someone sitting directly in the center.

Speaker Matching

Thus far, all of our checks have been made separately for each speaker. There's also the question of how well the speakers work together.

For the front speakers, this test requires that you put on programming that moves across the screen. The movie that's always suggested for this is *Top Gun*, which has planes taking off across the screen from one side to the other.

With all three (or two) front speakers on (but with surround speakers unplugged), run the tape to the point where a plane takes off. You should hear the sound move across the front of the room. You should *not* hear a break when it moves from one speaker to the next. The flow should be smooth.

If you hear the sound change pitch when it transitions from one speaker to another, the speakers are mismatched. They aren't putting forth the same harmonics or volume, so the sound of one is different than the sound of another. If you have trouble with this concept, think of bells. A bell made of glass and one made of silver may ring the same note, yet sound distinctively different. Similarly, two

speakers may play the same music or tones, yet sound distinctively different. Ideally, you want speakers that sound the same.

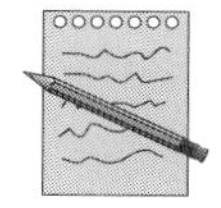

Note: Be sure you try this with your center speaker, too. It's not enough to have just the satellite (left/right) speakers matched. The center speaker must be matched to them as well.

Self-Powered Speakers

From the above discussion, it should be clear that the center front speaker should be as powerful as the two satellite speakers. Additionally, if you use a subwoofer, it too should be of adequate power.

However, if your amplifier/receiver offers subwoofer and center speaker control via a pre-amp, but does not supply power amplification, all is not lost. You can use *self-powered speakers* (as shown in Figure 7.4). These are available from a variety of manufacturers, and can provide just the power you need where you need it. (Self-powered satellite speakers are available as well.)

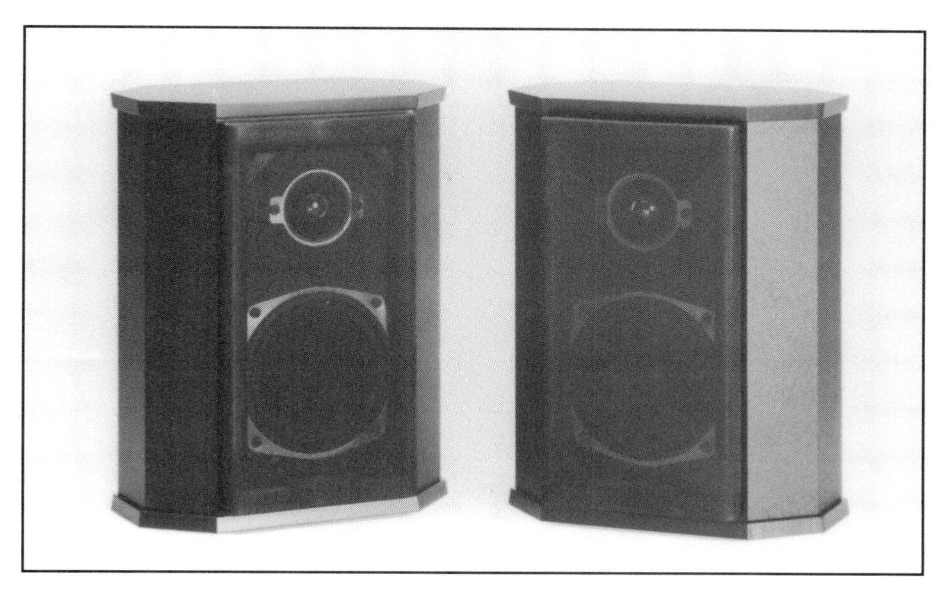

Figure 7.4. Jasco Model 418 self-powered speakers.

Hiding Speakers

The proper maxim with speakers is that they should be heard, but not seen. You want the sound, but you don't want the sight of ungainly big speakers. There are a variety of solutions to this problem.

One alternative, if you haven't already considered it, is to use a subwoofer. Because the subwoofer's sound is nondirectional, it can be discretely placed almost anywhere toward the front of the room. This then allows for much smaller and less-obtrusive satellite speakers.

Another alternative is to get very small speakers. Bose pioneered this concept with amazingly small speakers that produce large sounds (shown in Figure 7.5). Polk and other speaker manufacturers offer similar small yet high-performance speakers.

Figure 7.5. The tiny Bose Acoustimass 7 speaker system includes three dual-cube arrays and a bass module. Note the pen in the foreground for a perspective on size.

These small speakers can be concealed on shelves or as part of other home furnishings. By using them, you can avoid having to make your speakers the central element of your room design.

Yet another method is to recess the speakers into the wall (as shown in Figure 7.6). A number of manufacturers, including Boston Acoustics, make high-quality wall speakers. Space for these is cut into the wall and the speakers are recessed. Their cloth covering is unobtrusive and can be changed to match the decor of the room. (Note: Keep in mind that any cloth covering a speaker should be acoustically transparent - almost all manufacturers take care of this, but it's a consideration if you replace the cloth. Also, putting the speaker behind things to conceal it will muffle its output.)

See Chapter 10, "Furniture and Lighting that Make the Difference," for much more information on optimizing the appearance of the room.

The Many Kinds of Speakers

Thus far we've been talking about speakers as if they were all built the same way. With today's technology, however, there are a variety of different ways to build speakers. Some methods produce sounds better in the higher ranges, others in the lower. Many speaker systems incorporate several different types of speakers.

As your price horizon goes up, you'll increasingly run into exotic equipment—some of which doesn't even look like that which traditionally might be called a speaker. Here are some of the different types of speakers available today:

> *Dynamic loudspeaker*—This is what everyone thinks of when they visualize a speaker. There's a cone of some paper/cloth-type material with a heavy magnet in the center. The largest of these are used to handle deeper sounds and are called *woofers* or *subwoofers*. Smaller dynamic loudspeakers, often only three or four inches across, handle mid-range sound. *Tweeters* are tiny versions of this speaker for the very highest notes. An entire system can be built of dynamic speakers.
>
> *Planar speakers*—Instead of a cone found in the dynamic loudspeaker, here there's a thin sheet of plastic. This *diaphragm* has wire tracks glued into it which carry the signal. Planar dynamic speakers typically are composed of panels, instead of boxes, and they are thin; often, only an inch or two thick.
>
> *Electrostatic (transparent) speakers*—Similar to planar speakers, here the coil (the electromagnet powering the speaker) is formed in a grid at the edges

of the flat plastic diaphragm. These speakers are thin and very lightweight.
Ribbon—This is an unique type of speaker. Here a ribbon of metal is suspended in a strong magnet. The ribbon vibrates when the signal is passed through it, producing sound. Sometimes the ribbon is plastic with metal embedded in it.

Figure 7.6. Boston Acoustics 360 series in-wall speakers.

For the last three speaker types, the range varies depending on the construction. In high-end (read "expensive") speaker systems, several different technologies may be incorporated to produce sound.

Selecting Your Speakers

There are dozens of speaker manufacturers today, most of whom produce excellent quality units. Many, by the way, are assembled in the United States, as opposed to the electronics of your system most of which is produced abroad. However, the key to getting just the right speaker is to try the speaker out.

For this, I suggest you shop around. Stores from high-end specialty shops to discount chains usually carry several brands and many are set up in home-theater environments. Check out the speakers for both quality and price. You can expect to pay a suggested retail price of anywhere from around $500 for a group of five speakers (add several hundred dollars more for a subwoofer six speaker system) to $5,000 and much, much more for a top-of-the-line speaker system. (See Chapter 3, "Building A System for $1,500 to $15,000," for more information on pricing.)

The speakers you ultimately get will depend on three separate factors:

Your pocketbook
Your tastes in sound
Your desire to upgrade later on

Whatever you do decide, remember that money spent on your speakers is well spent. They produce the sound you will listen to. You can't go wrong getting the highest quality you can afford.

8 CHAPTER

Video Ware

No discussion of home theater would be complete without a brief look at the electronic media that you can use on your system. After all, the best A/V system in the world isn't going to do you any good unless you have a movie to run on it.

Locating Movies

The basic video software today is the VHS tape and the laserdisc. As noted elsewhere in this book, films shot in Dolby Stereo usually have Dolby Surround when they're reproduced on VHS tapes and laserdiscs. If it says Dolby Stereo on the box, it will play on your Dolby Surround Pro-Logic A/V receiver. But where do you find just what you want?

The answer for most of us is to jog on down to our local Blockbuster Video or other video rental store and rent the movie we want to see. The problem here is that the laserdisc selection often is limited, if even available. And the movie selection may, itself, be limited. Further, for the true aficionado, it's often not enough to rent. The goal is to buy the software (tape or disc) so as to have it in pristine condition for playback.

If you're looking to buy A/V software, there are a number of resources available to you. Some will sell by mail order, while others will help you find just what you're looking for.

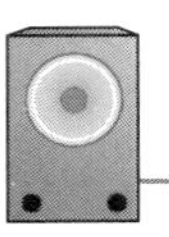

Dolby Laboratories Information

The Dolby A/V Forum is now available on America Online, a popular computer online service (via modem) for users of MS/DOS, Apple, and Mac platforms. Forum subscribers can browse through a variety of information on Dolby technology, as well as download lists of valuable information. Some of the downloadable data includes a list of the more than 3,600 Dolby Stereo films released to date.

Further, if you have information on Dolby, you can leave questions on a message board, and, as time permits, these will be answered by Dolby Labs staff. You can also communicate with other subscribers to the forum answering and asking questions of each other.

A unique feature of the Forum is the "Manufacturer's Corner." This is a message board where users can communicate with manufacturers of consumer electronics equipment and get feedback on questions and other issues.

For more information on the Dolby A/V Forum, contact America Online—800/827-6364, extension 5754. You'll get a free start-up kit and can subscribe.

Databases

You also can obtain databases for your computer that will include a wide variety of information on movies. For example, the FLICK! Film Review Library for the DOS platform includes over 30,000 movies. It gives information on who's in the cast, who directed, what kind of reviews the movie received, what its rating was, and much more.

As with other databases, you can search, sort, and edit the data to come up with just the information you want. The most recent price was under $60 from Villa Crespo Software, 1725 McGovern St., Highland Park, Illinois 60035—708/433-0500.

Laserdisc Listings

To get a complete listing of laserdiscs that are available, consider *Laser Video File*. This is a magazine in booklet form put out twice a year by New Visions Inc. It includes a photo of the cover, price, sound, color, rating, time, and more information on numerous films. In addition, there are occasional supplements, such as *Widescreen Movies On Laserdisc* ($2.95).

The periodical is available at some distributors such as Tower Records. It's also available by subscription for around $13 a year. Write to *Laser Video File*, PO Box 828, Westwood, NJ 07675—201/712-9500.

Dolby Discs in Surround

In addition to movies, a wide variety of compact discs have been recorded in Dolby Surround so that you can get a Surround listening experience. Here's a list of the various discs that were available as of last year:

CONCORD JAZZ

Frank Wess
Entrenous
CCD-4456

Mel Torme
Night at the Concord Pavilion
CCD-4437

Mel Torme/George Shearing
"Do" World War II
CCD-4471

Poncho Sanchez
A Night at Kimball's East
CCD-4472

Various Artists
Live at the 1990 Concord Jazz Festival—Volume 1
CCD-4451

Various Artists
Live at the 1990 Concord Jazz Festival—Volume 2
CCD-4452

Various Artists
Live at the 1990 Concord Jazz Festival—Volume 3
CCD-4454

PRO ARTE

Columbus Symphony Orchestra/Badea
Bartok: Miraculous Mandarin, etc.
CDD 535

Copenhagen Philharmonic/Entremont
St. Saens: Organ Symphony
CDD 534

Dallas Symphony Orchestra/Mata
Spanish Orchestral Music
CDD 536

Dallas Symphony Orchestra/Mata
Stravinsky: Petrouchka
CDD 537

Peter Nero and the Columbus Symphony
Anything But Lonely
CDD 522

Helsinki Philharmonic/Comissiona
Classical Storm Music
CDD 530

Rochester Philharmonic/Elder
Gilbert & Sullivan Overtures
CDD 533

San Diego Symphony Orchestra/Schifrin
Those Fabulous Hollywood Marches
CDD 504

San Diego Symphony Orchestra/Talmi
Gliere: Symphony No. 3
CDD 538

Various American Orchestras
A Broadway Spectacular
CDD 529

Various American Orchestras
Orchestra Spectaculars of Classical Music
CDD 528

Various Orchestras
The American Home Video Classical Album
CDD 529

Vienna C.O./Entremont
Mozart: Symphonies 40 and 41
CDD 531

Vienna C.O./Entremont
Mozart: Eine Kleine Nachtmusik, etc.
CDD 532

RCA VICTOR

Arthur Fiedler/The Boston Pops
Motion Picture Classics, Volume One
60392-2/4-RG

Arthur Fiedler/The Boston Pops
Motion Picture Classics, Volume Two
60393-2/4-RG

Charles Gerhardt/The National Philharmonic
Now Voyager/The Classic Film Scores Of Max Steiner
0136-2/4-RG

Charles Gerhardt/The National Philharmonic
Captain From Castile/The Classic Film Scores Of Alfred Newman
0184-2/4-RG

Charles Gerhardt/The National Philharmonic
Casablanca, Classic Film Scores for Humphrey Bogart
0422-2/4-RG

Charles Gerhardt/The National Philharmonic
Gone With The Wind/The Classic Film Score by Max Steiner
0452-2/4-RG

Charles Gerhardt/The National Philharmonic
Citizen Kane/The Classic Film Scores of Bernard Herrmann
0707-2/4-RG

Charles Gerhardt/The National Philharmonic
Sunset Boulevard/The Classic Film Scores of Franz Waxman
0708-2/4-RG

Charles Gerhardt/The National Philharmonic
Classic Film Scores for Bette Davis
0813-2/4-RG

Charles Gerhardt/The National Philharmonic
Elizabeth & Essex/The Classic Film Scores of Erich Wolfgang Korngold
0815-2/4-RG

Charles Gerhardt/The National Philharmonic
Spellbound/The Classic Film Scores of Miklos Rozsa
0911-2/4-RG

Charles Gerhardt/The National Philharmonic
Captain Blood/Classic Film Scores for Errol Flynn
0912-2/4-RG

Charles Gerhardt/The National Philharmonic
Lost Horizon/The Classic Film Scores of Dimitri Tiomkin
1669-2/4-RG

Charles Gerhardt/The National Philharmonic
The Spectacular World of The Classics Film Scores/Sampler
2792-2/4-RG

Charles Gerhardt/The National Philharmonic
Close Encounters/Star Wars, John Williams' Classic Film Scores
2698-2/4-RG

Charles Gerhardt/The National Philharmonic
Star Wars: Return of the Jedi, John Williams' Classic Film Score
60767-2/4-RG

Charles Gerhardt/The National Philharmonic
The Sea Hawk/The Classic Film Scores of Erich Wolfgang Korngold
60863-2/4-RG

David Raksin/The New Philharmonia Orchestra
Laura/Forever Amber/The Bad & The Beautiful
1490-2/4-RG

Henry Mancini/The Mancini Pops Orchestra
Mancini In Surround—Mostly Monsters, Murders & Mysteries
60471-2/4-RC

Henry Mancini/The Mancini Pops Orchestra
Mancini's "Monster" Hits (Collector's edition; glows in the dark)
60577-2/4-RV

Henry Mancini/The Mancini Pops Orchestra
Cinema Italiano—Music of Ennio Morricone & Nino Rosa
60706-2/4-RC

Henry Mancini/The Mancini Pops Orchestra
The Pink Panther & Other Hits—newly remixed original recordings
7863-55938-2/4

Isao Tomita
Kosmos
2616-2/4-RG

Isao Tomita
The Tomita Planets (Holst)
60518-2/4-RG

Isao Tomita
Pictures at an Exhibition (Mussorgsky)
60576-2/4-RG

Isao Tomita
Firebird (Stravinsky)
60578-2/4-RG

Isao Tomita
Snowflakes Are Dancing (Debussy)
60579-2/4-RG

London Cast Recording
Into The Woods—Stephen Sondheim/James Lapine
60752-2/4-RC

Original Soundtrack Recording
Altered States, John Corigliano
3983-2/4-RG

Various Artists
Silver Screen Classics—Four classic film scores (collector's edition)
60763-2/4-RG

Various Artists
The Home Video Album
60354-2/4/9-RC

Broadcast in Dolby Surround

In addition to tapes and discs, many TV shows are broadcast in Dolby Surround. These shows, as noted earlier, will say on the credits, "Recorded in Dolby Stereo." That means that you can use your Dolby Surround system when listening to them (as long as your local cable station provides stereo to your home). Television shows that have been recorded in Dolby Surround over the past few years include the following:

The Arsenio Hall Show—syndicated
Austin City Limits—PBS
Beverly Hills, 90210—FOX
CBS Sports—CBS
Donohue—syndicated
Dinosaurs—ABC
Dr. Quinn, Medicine Woman—CBS
Evening Shade—CBS
Jack's Place—ABC
Late Night/Late Show with David Letterman—NBC/CBS
Mad About You—NBC
Melrose Place—FOX
NARAX/Grammy Award Show, 1993—CBS
Nashville Now—TNN
The Nature Series—PBS
Northern Exposure—CBS
Raven—CBS
Rin Tin Tin, K9 Cop—FAM
The Simpsons—FOX
Space Rangers—CBS
Star Trek: The Next Generation—syndicated
Tales From The Crypt—HBO
Texas Connection—TNN
The Tonight Show—NBC
The Young Indiana Jones Chronicles—ABC
Zorro—FAM

...and others.

Other Resources

In addition to information on videos, there are a wide variety of booklets, magazines and other materials published on video software. For example, Philips puts out a brochure listing all software available for its CD-I player. For information call 800/824-2567.

For games, there are many publications that cater to Nintendo, Sega, Acclaim, and other manufacturers of both hardware and software. These are widely available on newsstands and within them list other resources.

For overall information on the video field the following publications may prove useful. (As noted, some are "trade publications," meaning that they are intended for distribution within the field. However, almost all will allow you to subscribe as a user.)

A/V Video
Montage Publishing
701 Westchester Ave.
White Plains, NY 10604

Audio Video International
PO Box 384
Winchester, MA 01890-9958

Camcorder Magazine
4880 Market St.
Ventura, CA 93003

NewMedia
PO Box 1771
Riverton, NJ 08077-9771

Stereo Review
PO 55627
Boulder, CO 80322-5627

Stereophile
PO Box 52977
Boulder, CO 80322-2977

VIDEO
PO Box 52255
Boulder, CO 80321-2255

9 CHAPTER

Putting It All Together

One of the great mysteries of the world for many of us is how to hook up all the wires at the back of our home-theater system. Indeed, I suspect that some people hesitate to build a system just out of fear of messing up the connections. How would it look, after all, if you bought all the components, assembled everything in the room, and then had to hire an engineer to connect it?!

Rest assured that hooking it all together is not only possible, but downright easy when you know how. In this chapter we're going to examine some of the tricks of a quick hook-up so you won't be lost (or have to call an electrical engineer). We'll also look at some handy extras, such as universal remotes, that can make controlling your system a whole lot easier.

Planning Ahead

The key to simple hook-ups is to plan ahead. I always suggest putting it on paper. Draw a simple diagram of where everything is going to go including your wiring runs. Then label all wires at each end. You can use a small piece of "Post It" paper with some Scotch tape to accomplish this. (This is probably the best single tip you'll ever receive!)

As soon as you begin a diagram, you'll realize that some things are obvious. All of your equipment will go in the same room. Your television and most of your speakers will be clumped in the front against one wall. The surround speakers will be in the back and/or sides. The A/V receiver/amplifier, VCR, CD player, and other electronics probably will be all together, though not necessarily next to the TV.

So, what have we got?

We've got a center and extremities. Think of it as a wheel with the hub at the center (the electronics) and the spokes (the cables) leading out to the extremities (the speakers and TV). Figure 9.1 shows a typical wiring diagram.

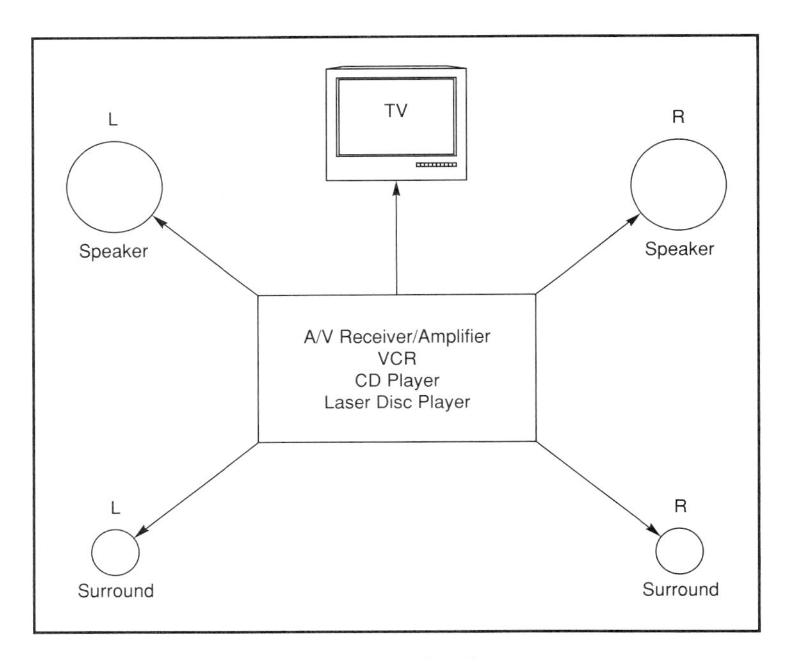

Figure 9.1. A "wheel" diagram of a basic hook-up.

Wiring Runs

Once you have your basic organization, figure out where everything will go in the room. Plan your wiring runs. These should be as simple as possible. Here are a couple of tips:

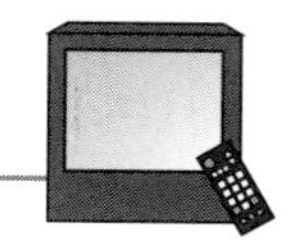

- Keep speaker wires out of sight. You can accomplish this by running them under the carpet (be sure to avoid high-traffic areas as this will break down the wires and reduce their ability to transmit a clean electrical signal); tacking them to a baseboard; or running them through a wall, under the house through the basement (or up into an attic, or outside the house) and then in again at the proper location. The latter may seem drastic, but I've done it. I painted the wires and they're barely visible outside. In a basement or attic, you probably won't care if they show. If your speakers are up high near the ceiling, try to run the wires up in a corner. You can conceal them with drapes or other wall hangings.

- Use high-quality wiring for your speakers. (See the section titled "A Special Note on Wiring" in this chapter.) There are a variety of types, but the most practical is multistrand copper that's usually far more heavy-duty than you might think necessary. Some companies sell gold- or silver-plated wire, and many aficionados swear by it. Gold and silver definitely are better conductors of electricity than copper. However, in a wiring run of 20 feet or so, the difference in power loss at all frequencies between copper and a precious metal probably is not worth the extra cost. This is particularly the case with thick, multistrand, copper wire, so you may be paying a lot for only a little gain. Gold-plated connectors at the ends of wires, however, could make a bigger difference. You may want to consider these.

- The electronics must be near an electrical outlet. Dragging a heavy AC cord across the floor is a definite no-no. Ideally, you'll have at least one 20-amp dedicated receptacle. *Dedicated* means that there's nothing else, such as a washing machine or refrigerator, connected to the circuit. Having these other items on the circuit could mean they'd cycle on in the middle of *Top Gun*, overload the wiring, trip the circuit breakers, and close you down in the middle of a dramatic scene. Unfortunately, many inside receptacles are only on 15-amp circuits, each of which probably also services many other receptacles. You may have to add a 20-amp circuit breaker to the main panel box and run No. 12 solid wires to a new receptacle you install for your electronics.

> **Warning:** If you're not competent with handling electrical wiring in the home, this will be a job for an electrician.

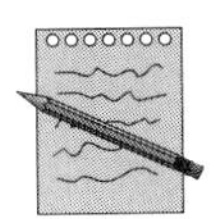

Note: You can try using two or more different receptacles, if they are on separate circuits. The trouble is that in most houses, all the receptacles in a room are on the same circuit.

- Plan your room lighting at the same time as your equipment hook-ups. If you're going to install track lighting, extra receptacles, or switches, do it all at the same time. This will save you money and aggravation.

Warning: Be sure to check and comply with all local building and electrical codes prior to any electrical modification.

A Special Note on Wiring

The type of wiring you choose is important. It's so important, in fact, that the Home THX people have licensed Monster Cable to create special Home THX wiring. Monster Cable produces a cable with many separate wires inside, each properly labeled, each having the right diameter and makeup. You don't have to use this wiring—but whatever wires you use must be big enough to handle the load you place on them in order to get the proper response from your speakers.

Organizing Equipment

It's fine to put your A/V receiver/amplifier, VCR, CD player, and other equipment in one centralized location as long as they are in a cabinet with plenty of air circulation. (See Chapter 10, "Furniture and Lighting that Make the Difference," for ideas here.)

Avoid stacking equipment on top of each other, even if there appears to be some air space in between. Even with circuit boards and chips, consumer electronics pieces generate large amounts of heat and, to avoid unwanted noise, usually do not have their own cooling fans. That means they count on external air flow for cooling. Allow at least an inch between components. This will give you better results and increase the lifespan of the units.

It really doesn't matter which equipment goes on top and which on the bottom. What does matter is that everything is as convenient as possible.

Typical Dolby Surround Hook-Up

If you're going to put together a basic Dolby Surround system, you will have a minimum of four speakers (front left/right and two surround), a television, perhaps a VCR and CD, a Dolby Surround decoder, an amplifier/receiver, and possibly a mixer, as shown in Figure 9.2. (*Mixers* are discussed later in this chapter under "Hooking Up an Audio/Video Mixer.")

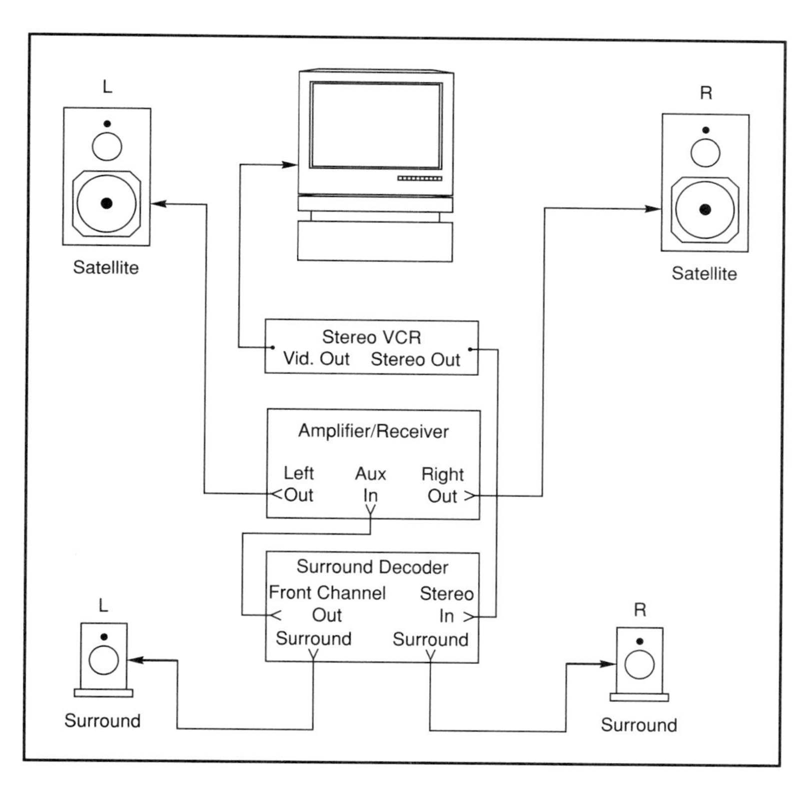

Figure 9.2. Basic Dolby Surround hook-up using an existing receiver/amplifier.

Note the wiring in Figure 9.2. It assumes that you *are not* using an A/V receiver/amplifier, but instead are using your existing receiver/amplifier and a Dolby Surround decoder. To be sure you understand it, here's a wire-by-wire explanation.

- *Surround-speaker wires*—The surround speakers form a monaural channel. However, there are separate wires running from your Dolby Surround decoder to each of these speakers. Be sure you connect the surround speakers only to the decoder. Your regular (existing) receiver/amplifier probably will not have connections for them.

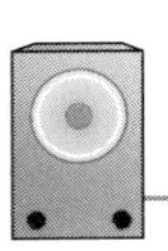

- *Front-speaker wires*—These should be connected to your existing stereo amplifier/receiver in the usual way. Be careful to hook up the left and right speakers to the left and right connections so that the balance control in the front of your unit will work properly.
- *Television wires*—You'll be using your television set partly as a monitor for your VCR and partly as a receiver. You will want to be able to play tapes from your VCR onto your TV or receive broadcasts. Therefore, either the composite signal cable, the RCA connectors for the separate audio/video signals, or the Y/C wires (independent luminance and chrominance on S-VHS sets) should go from your VCR's VIDEO OUTPUT to your television set's VIDEO INPUT. This way, you can either use your TV independent of your VCR (when the VCR is switched off) or play the VCR and have the picture sent to your TV. Your VCR has a TV/VCR switch, which allows you to bypass the VCR with a broadcast signal when you're taping video off the air.
- *VCR wires*—As already noted, the VIDEO OUTPUT will go directly to your television set. The VIDEO INPUT may come either from cable or antenna or from your Dolby Surround decoder. (We'll have more to say about video input from broadcast through a cable box later in this chapter, under "The Cable-Box Conundrum.") If your VCR has stereo, you will connect your STEREO OUTPUT to the decoder's STEREO INPUT.
- *Decoder wires*—Your Dolby Surround decoder primarily will be used to send out the surround-channel sound. It will have at least three outputs and one input. Connect the SURROUND OUTPUTs to the surround speakers. Connect the FRONT CHANNEL OUTPUTs to the STEREO INPUTs on your existing amplifier/receiver. (You may have to use inputs marked "auxiliary" or "tape" for this. Experiment a bit until you find some that work.) Connect your STEREO AUDIO INPUT from your VCR's stereo audio output. If your decoder has a CABLE INPUT, connect it from your cable and send your decoder's VIDEO OUTPUT to your VCR.
- *Stereo receiver/amplifier wires*—The wires from your decoder will go to your AUXILIARY STEREO INPUTs. The wires from your SPEAKER OUTPUTs will go to your front speakers.

Typical Dolby Surround Pro-Logic Hook-Up

For this, the assumption is made that you'll be using an A/V receiver/amplifier that can handle many video and audio inputs and outputs. For example, it will accomodate the various video sources you will use, including broadcast, VCR, and laserdisc player. It also will handle various audio inputs, including cassette player, CD player, and record player. (The Dolby Surround Pro-Logic A/V receiver/amplifier is discussed in more detail in the next section.)

In addition, the assumption is that you will also be using a six-speaker system (left/front/center, two surround, and a subwoofer). Figure 9.3 shows a typical Dolby Surround Pro-Logic hook-up. (Note: Be sure the polarity is the same for all speakers, or else you may detect a weaker or strange sound in one. Usually speakers as well as amplifiers have red and black connectors. Be sure the same wire coming out of your electronics goes to all red or to all black connectors—don't mix them.)

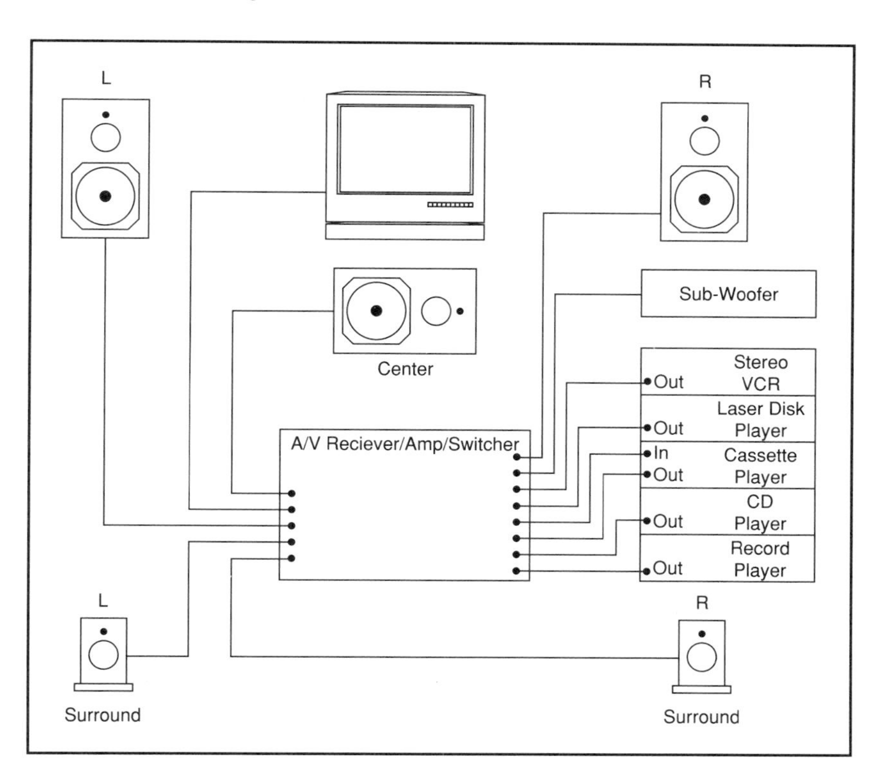

Figure 9.3. Dolby Surround Pro-Logic home-theater wiring hook-up.

Note the wiring shown in Figure 9.3. It assumes that you buy a Dolby Surround Pro-Logic A/V receiver/amplifier and that you have a variety of external sources, including a VCR, laserdisc, cassette, CD player, and record player. The hook-up is more complex, but still logical. Let's take it a step at a time:

- *Surround-speaker wires*—The surround speakers form a monaural channel. However, there are separate wires running from your A/V receiver/amplifier to each of these speakers. Be sure you connect the surround speakers only to the SURROUND OUTPUT. Normally, the front output will have much more power—and, if improperly connected, it could blow out your surround speakers.
- *Front-speaker wires*—These should be connected to your A/V receiver/amplifier's FRONT CHANNEL OUTPUTs. Be careful to hook up the left and right speakers to the left and right connections so that the balance control in the front of your unit will work properly.
- *Subwoofer wires*—These connect the SUBWOOFER OUTPUT on your A/V receiver/amplifier.
- *Center-speaker wires*—These connect to the CENTER OUTPUT on your A/V receiver/amplifier.
- *Television wires*—You'll be using your television set partly as a monitor for your VCR and laserdisc and partly to receive broadcast. You will want to be able to play tapes from your VCR onto your TV or get a live signal. Therefore, you'll be using your A/V receiver/amplifier as a switcher. Your television's wires will connect directly to the A/V receiver/amplifier's VIDEO OUTPUT.
- *CR wires*—As already noted, the A/V receiver/amplifier will be used as a switcher. Therefore, your VCR's VIDEO OUTPUT will go directly to the A/V receiver/amplifier's VIDEO INPUT, and the VCR's STEREO OUTPUT will go directly to the A/V receiver/amplifier's STEREO INPUT.
- *Laserdisc player*—This will be hooked up exactly the same as the VCR. The VIDEO and STEREO OUTPUTs will go directly to the A/V receiver/amplifier's VIDEO and STEREO INPUTs.
- *CD, cassette, and record player*—All your audio sources will connect directly to the A/V receiver/amplifier's AUDIO INPUTs.

Dolby Surround Pro-Logic A/V Receiver/Amplifier

Your Dolby Surround Pro-Logic A/V receiver/amplifier will be used as a mixer for both audio and video sources. Since most such units have many jacks in the rear to accommodate this use, you shouldn't have much trouble. (When you buy your A/V receiver/amplifier, though, it's a good idea to check in the back to be sure it has a sufficient number of audio/video jacks.)

To reiterate, here are the connections to be made:

1. Connect the VIDEO OUTPUT to your TV and VCR.
2. Connect the VIDEO INPUTs to cable and your VCR and laserdisc.
3. Connect AUDIO OUTPUTs to your cassette-tape player.
4. Connect the AUDIO INPUTs to your casette-tape player, CD player, and record player.
5. Connect all SPEAKER OUTPUTs to their appropriate speakers. You may need a self-amplified subwoofer or center speaker, depending on your A/V receiver/amplifier.

The Cable-Box Conundrum

Thus far we've taken a straightforward look at hooking up your home entertainment system components. Now, let's add another, sometimes confusing, element: the cable box.

One programming source for you is most certainly going to be broadcast TV. This is especially the case as more and more shows are recorded and broadcast in Dolby Stereo sound. Why not use your system to its maximum capabilities as long as the signal is there?

The answer is that the signal may or may not be there. Check with your local cable company to see if their equipment is transmitting the signal as originally recorded. Some cable companies have the capacity to send stereo over their lines, others do not.

Assuming that your cable company does send in stereo, the next question is: Where do you connect the cable box? Do you put it in front of everything else so that the first thing the cable goes through is the box? Or do you put it between your A/V receiver/amplifier and your TV? What about the case in which you're using a stereo VCR? Do you put it before the VCR or after?

Figure 9.4 shows a wiring hook-up from outside the cable system.

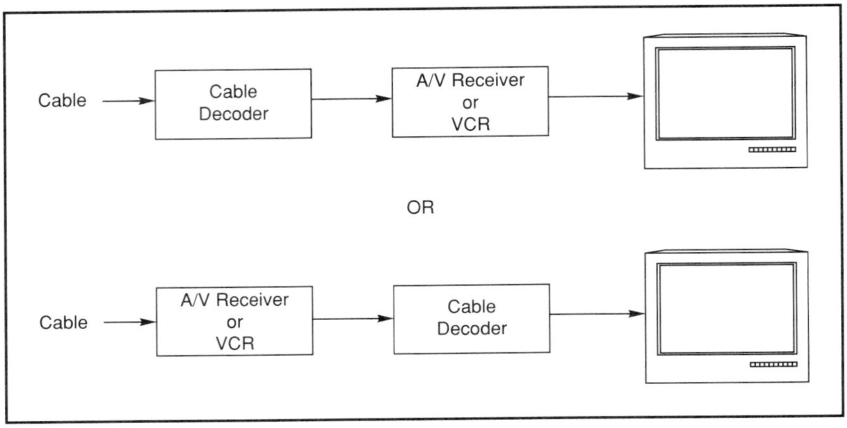

Cable → Cable Decoder → A/V Receiver or VCR →

OR

Cable → A/V Receiver or VCR → Cable Decoder →

Figure 9.4. Cable-wiring hook-up options.

Most cable companies want you to put their box on top of the TV and connect directly to the TV. That's to be sure that you don't strip off an extra signal that you could send elsewhere in the house. (Some cable companies charge for each outlet.)

In order to get a stereo signal, however, you may need to put the cable box in front of your A/V receiver or before it goes into your VCR. (Note, however, that your cable company may charge a separate fee for taking FM stereo transmission off of its line.)

The reason to do it this way is that you may subscribe to one or more of the premium channels. You need the cable to decode the signal. However, do you need the "black box" from the cable company to decode the audio portion? It depends on your cable company's equipment and their method of scrambling.

The only real solution is to try it several different ways. Keep in mind, however, that cable-company black boxes are not renowned for their A/V quality. Every time the signal goes through one you can be sure you're going to get some degradation. To overcome this, some cable companies boost their video signal—which may only result in a different kind of distortion on your set.

Satellite Hook-Ups

If you're using your own home-satellite system, you have an additional concern: how to handle the satellite signal.

Today you will undoubtedly be using an *Integrated Receiver Descrambler*, or IRD. This device handles the actuator on your dish, which moves it as well as receives signals in both C and Ku bands. In addition, it descrambles signals from premium channels (such as HBO, Showtime, Cinemax, and so on). And it delivers that signal in stereo with Dolby Surround, as long as it's broadcast in that manner.

IRDs are high-quality instruments and you can have perfect confidence in putting them first in line before your VCR or A/V receiver/amplifier. In any event, you have no choice. You can't get any kind of signal from your dish unless the IRD goes first!

Although we've briefly discussed a satellite hook-up in this chapter on putting your system together, Chapter 13, "Hooking Up to a Satellite," tells you much more about the topic of satellites and their technology.

A-B Switches

Many people who have satellite systems also have either a conventional TV antenna or a cable hook-up. This is because they want to receive broadcasts such as local news and shows from nearby television stations. (Local stations send their signal via broadcast or cable, but *not* via satellite.) Hence, if you have a dish, you're also likely to have two signals coming in—one from cable/antenna and one from satellite. If you have an A/V receiver with multiple inputs, you can handle this easily by plugging the satellite signal into one input and the cable/antenna input into another using the unit as a mixer.

If, however, you only have a VCR stereo system, you can easily accomplish the same thing via an *A-B switch*, shown in Figure 9.5.

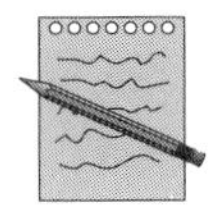

> **Note:** You're not likely to have to worry about a cable "black box" on the line for handling the audio on movies and broadcast shows as you'll be picking up all premium channels as well as the networks and descrambling them with far superior results with your IRD.

An A-B switch is a double-isolated throw device that allows you to change between two inputs. One direction gives you cable/antenna, the other gives you satellite.

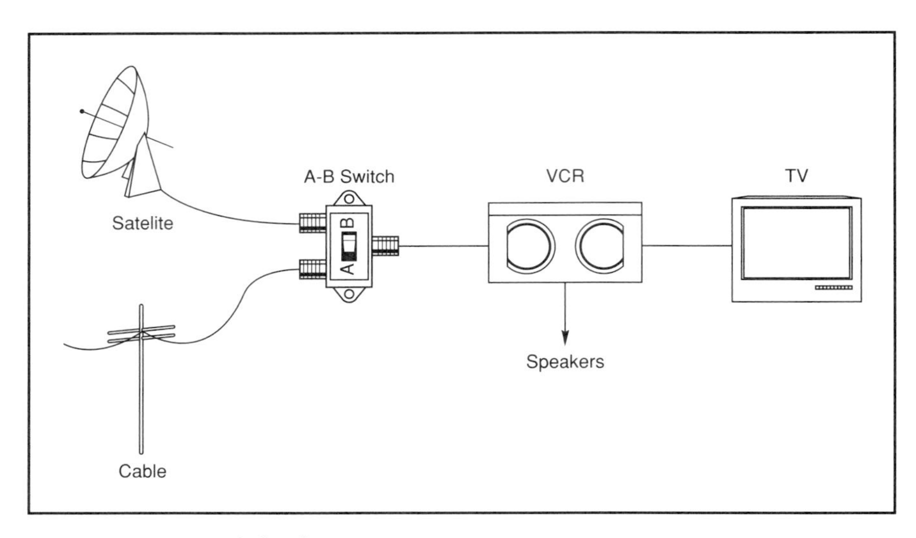

Figure 9.5. A-B switch hook-up.

Hooking Up an Audio/Video Mixer

This brings us to those who have multiple video and audio inputs and outputs and don't have the means to handle them. Ideally, you would buy an A/V receiver/amplifier with enough video and audio inputs and outputs to handle all your needs. But, perhaps you have more than the usual number of video and audio sources. In that case, you can purchase a *mixer*, as shown in Figure 9.6. These are available for as little as a couple of hundred dollars on up into the thousands. The expensive models have dials to show signal strength and some amplification features. Be sure you get a high-quality unit, preferably with connections coated with precious metal so the mixer doesn't significantly degrade the signals going through it.

A mixer is really just a bunch of A-B switches all run together. It allows you to switch between any number of sources and outputs. Usually the only hard thing about it is reading the names of the sources on the jacks on the back of the unit. Typically they're written in black on black, and you need a powerful light and perhaps a magnifier to discriminate between them. In any event, the only reason you need to read them is so the sources you choose will coincide with their names on the front of the unit. (Some audio sources, such as tape, may require different levels of amplification and, consequently, must be hooked up correctly.)

INPUTS
Satellite Feed
Cable
Antenna
VCR
Laser Disc
Camcorder
CD Player
Cassette Player
Record Player
A/V Receiver/Amplifier

Mixer

OUTPUTS
A/V Receiver/Amplifier
VCR
TV
Cassette Recorder

Figure 9.6. Typical wiring for a mixer using multiple inputs and outputs.

Universal Remote

This brings us to one last matter which has become increasingly important as more and more components have been added to home-theater systems. I'm speaking of the remote control(s), often called simply the *remote*(s).

I always recommend a remote for every component you have. You need the remote so you can sit in the appropriate place in your room and make all the necessary adjustments to your system.

However, this means that you're going to end up with a lapful of remotes. You could conceivably have a separate remote for your A/V receiver/amplifier, TV, VCR, CD player, laserdisc player, satellite receiver, cable box, and so on. It's not inconceivable that you could end up with a dozen different remotes and spend a good portion of your time trying to find the right one.

Manufacturers are not unaware of this problem and have been working diligently on solutions. One solution they particularly like is a remote that works all of the elements of your system, provided they are all from the same manufacturer. Thus, this universal-remote-that-is-not-universal will work your XYZ TV, A/V receiver/amplifier, and VCR at the least, and may also work other pieces of equipment.

However, as we've seen in this book, it's often less expensive, and produces better-quality results, to mix and match many components. What do you do about too many remotes, then?

The answer is a true universal remote. These are just now coming onto the market at a reasonable price. Most are using a new chip that comes programmed with all of the infrared remote signals for the 14 major manufacturers of components. They also include a "learn" mode by which many signals not already coded into the device can be input.

One of the nicest I have seen is the Sony RM-V10, shown in Figure 9.7. It operates TVs, VCRs, and cable boxes and, with luck, some of your other A/V equipment as well. Best of all, its suggested retail price is only $19.95!

Figure 9.7. The Sony RM-V10 remote; a bargain at under $20.

Other, more-sophisticated universal remotes are coming onto the market which will operate almost all A/V equipment from major manufacturers. (Keep in mind that your equipment must receive an infrared signal for these remotes to work. If you have older equipment without a built-in remote, you're out of luck.)

What To Do When Nothing Works

If after this rather detailed explanation you find yourself sitting on the floor, with wires going everywhere and nothing seeming to work, relax. Take a break. Have a glass of wine or iced tea or whatever. Forget it for awhile. Later, when you're calmed down, come back and try this approach:

1. Label both ends of all wires in the manner described at the beginning of this chapter.

2. If not clearly labeled, label all outputs and inputs on all equipment in bright white.

3. Draw on a diagram where every wire goes. Compare it against the diagrams in this chapter to make sure you've got the inputs going in and the outputs going out.

4. Connect it up.

5. If it still doesn't work, it may be time to call a friend, electrician, or psychiatrist.

Good luck!

CHAPTER 10

Furniture and Lighting that Make the Difference

I once had a college professor who was a wine connoisseur. He could tell you the type and vintage of a wine from just a sip, and he loved to extol the virtues of the fermented grape. However, beyond wine, one thing that he said stuck with me; to paraphrase, "No matter what wine you are serving, it must be drunk from a beautiful, carefully made wine glass in order to properly enjoy it."

He was right. Packaging is an important part of the product. This concept also applies to home theater.

You can sit on an orange crate and listen to the sound coming from components spread over the floor and watch the video from your TV set balanced on another box. Or you can sit in a comfortable easy chair and have your equipment gracefully shelved, and your TV set placed in a beautiful cabinet. Technically, the sound and the sight are the same. However, I can guarantee you'll enjoy it much more in a well-furnished setting. In this chapter we're going to look a little at furnishings, visuals, cabinets, and lighting to see what you can build for yourself at minimal cost.

Cabinetry

Major furniture manufacturers (including Ethan Allen, Thomasville, and others) are well aware of the boom in home theater. They're producing high quality cabinets to house large screen TVs, A/V components, and even speakers. Typically these marvelous cabinets are made of exotic woods with polished finishes. And they are expensive. A typical A/V cabinet can easily run $5,000 or more. Add three of them together (for components, speakers, and storage) and you can have a $15,000 furniture wall before even buying the electronics!

However, a cabinet that holds your equipment does not have to be made of expensive woods and have a fine finish. From a utilitarian point of view, all that you need are racks to hold the components and a stand for the television set. And if you use a cabinet, you need a hole in the back into and out of which you can run the wiring.

If you're clever with your hands and your mind, you can design and build your own cabinets at a fraction of the cost of ready-made units. They probably won't look as exquisite as professional, quality-made furniture. But there's no reason they can't look good, and they certainly will do the job of holding your equipment just as well.

Your Basic Cabinet Needs

You're going to need three types of cabinets:

- Racks to hold the wide, long, thin electronic components
- Stands or shelves to hold the television and the speakers (if you want them to be integrated with the furniture)
- Storage space for your CDs, tapes, and, if you still have any, records

We're going to consider three separate methods of putting together cabinetry to handle all of these, all at nominal cost. Before doing so, however, here are some considerations regardless of what type of cabinets you have:

- *Cooling*—Because most components come with tiny legs, there's always a tendency to simply stack the components on top of each other without the use of shelves. Don't succumb to this temptation! The components all generate heat. Because we're dealing with sound, almost none of the components have fans to help dissipate that heat. The only way for the

heat to escape, therefore, is gravity. Heat rises and comes out those little slots usually cut into the tops of the component cases. Stacking one on top of the other restricts the air flow. The result is the component may not work as well, and you could significantly shorten its lifespan.

- *Power*—Your components are all going to need electrical power. One method is to send the power wires out the hole in the back of the cabinet and plug them into the wall. A better method is to buy a surge protector—something you most certainly will want to use to protect your expensive electronics. Plug it in and run it back into your cabinets. That way you won't have so many wires running out the back, and it'll be more convenient to plug things in.
- *Ease of change and repair*—Nothing lasts forever, and the placement of A/V components in your cabinetry is no exception. No sooner have you gotten it in than you'll probably want to change something. That means getting back there and unplugging all those wires, moving them around, and plugging them in again. If you build access into your cabinets at the beginning, it will make living with them much easier later on.

Wire Racks

Wire racks are made of heavy-gauge wire strung between metal supports. Some apparel stores use them to display sweaters and similar items. Typically, these shelves can be adjusted for height, the number of shelves can be changed, and different widths purchased. Also, they often come in white, which means that if you don't like the color, you can easily spray paint them any other color you want. In short, they are completely flexible, relatively light-weight, do a good job, look very modern and, perhaps most important of all, are very inexpensive. You should be able to find these at your local discount hardware store. A rack of wire shelves large enough to hold all your electronic components and a stand for your TV should cost well under $100. If you're using small satellite speakers, a small wire rack often can be used to give them height, again at nominal expense.

One point of concern about wire racks used as cabinets: they do conduct electricity. You may want to use some clear insulating plastic under your components, just in case they leak any current. On the good side, you can attach your FM radio leads directly to the rack. Sometimes they will make a wonderful antenna! (Of course, you can buy a good FM antenna for around just $25.)

What you need:

Tools required—None
Material Required—One or more ready-made wire shelves
Cost—Probably under $100

Wall Shelves

Another method that also looks good is to build wall shelves. If you build them correctly, you can have some shelves that are close together, others that are far apart. By adjusting the height and by buying different widths, you can have shelves that will hold all of your components, including TV and speakers.

The kind of shelf I speak of here is widely sold in hardware stores throughout the country under several labels, all of which fall under the heading of *modular shelving*. Modular shelving consists of metal brackets that you attach to a wall, shelf hooks that fit into the brackets, and prefinished shelves, which then rest on the hooks. Each piece is sold separately, so if you get a lot of shelves, the price can run up to a couple of hundred dollars. On the other hand, everything is prefinished, so once it's hung there's nothing else for you to do.

The shelves themselves are typically made of pressed wood with plastic "skins" to give them a solid, finished look. They often come in white (which can be painted), brown, gold, black, and occasionally in other colors. The brackets and hooks likewise come in a variety of colors.

These shelves are easy to assemble. If you haven't done it before, however, here are some tips:

- Because the shelves are going to be used to hold electronic components, and because those components usually are pretty heavy, it's best to be sure that the wall brackets are attached directly to studs. The instructions often suggest the use of wallboard bolts, which allow the brackets to be attached to wallboard or plaster. I would disregard those instructions and instead find the studs (the wood that the wallboard surface of your wall is nailed to) and use wood screws that go at least *one and one-half inches* into the wood. I have seen weight on shelves pull the brackets and their bolts right out of wallboard; it's an unpleasant sight, and a troublesome problem to fix. Also, get brackets that go up at least one foot and preferably two feet from your uppermost shelf. This gives you some extra holding power.

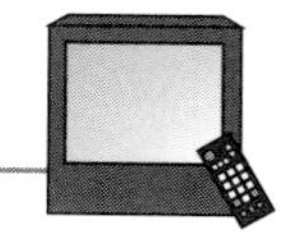

- Because the shelves are adjustable, you can place them as close together as you like. Allow at least an inch of air space between the top of one electronic component and the shelf above it. Two or three inches would be better. Remember, you have to allow room for the hot air inside the component to get out and dissipate, particularly if you plan to run the system for long periods of time.
- Watch out for deep components. The shelves typically come in widths of four, six, eight, and 12 inches with 16- and 18-inch widths sometimes available by special order. Because the components tend to be fairly deep, you're going to need the widest shelves. However, remember that the only thing holding the shelf up is the brackets, the screws, and the hooks. The wider the shelf, the greater the leverage pulling downward. Put a heavy component on a wide shelf and you've got enormous downward pressure. As a result, be sure to use at least three or more brackets per shelf. This helps spread the weight around and helps to ensure that you won't one day watch a shelf suddenly sag, bend, and let your A/V receiver/amplifier drop to the floor.

What you need:

Tools required—a screw driver, a nail, and a hammer (to help locate studs and drive a starter hole)

Materials Required—Brackets, hooks, and shelves. Figure you'll need a bracket every 16 inches and at least three brackets for every shelf. A typical wall-to-ceiling setup with seven shelves (spaced at different heights) by eight feet wide would require:

56 feet of shelving, typically in the form of 14 four-foot shelves

42 shelf hooks

six eight-foot brackets

Cost—Varies according to purchase location; probably under $200

Homemade Wood Cabinets

Finally, there's the homemade wood cabinet. This is a project for the do-it-yourselfer who has some tools out in the garage (and, hopefully, knows how to use them).

One of the big advantages of homemade cabinets is that you have the option (not usually available with the other types of shelving) of using flush-mounted shelves. In other words, you can build the shelving right into the wall so that, instead of protruding to the front, the lines are flush. This is very popular and has a very

pleasant look, although this design in terms of video enjoyment may not be optimal. (See the discussions on "Color" and "The Overall Appearance" in this chapter.)

I have seen homemade cabinets that rival the best that furniture manufacturers offer. And I've seen some that look "Jack built," slanted, wobbly, and ugly.

It's beyond the scope of this book to include plans for homemade wood cabinets (available in many different books at your local bookstore). However, when making them, keep the following in mind:

- Remember that the components are heavy. It's preferable to use grain wood or plywood (with veneer, if desired) that's at least five-eighths of an inch thick for the actual shelving with many supports. This is especially the case if the shelf is going to span more than two feet.
- Again, leave plenty of air space so that your equipment can keep cool.
- Don't forget to drill holes in the back for wiring. Make the holes big: one and one-half inches in diameter at a minimum, because you may need to push several plugs as well as wires through.
- Consider putting wheels on the cabinets so that you can move them around more easily once you have the components inside. It's often the case that only after you see everything filled out do you decide where you want it to go.
- Consider height before you begin to build. Most A/V component cabinets are fairly low. You may not want to go more than three or four feet tall. Also, remember that for Dolby Surround Pro-Logic, you're going to need to leave room for that center channel speaker (which, ideally, should go directly beneath the TV).

What you need:

Tools required—extensive woodworking tools
Material required—wood, screws, glue, metal fasteners
Cost—varies with the type of wood used, probably under $500

Chairs

The chairs and couches in the room should be placed facing the TV, in the center and toward the back. It really doesn't matter what kind you use, except that they should be comfortable. You're going to be watching movies that run two hours or

more. Remember the last time you were at the cinema and watched a long movie? Did the chair start poking you in the back? Did you squirm? Were you uncomfortable?

The only rule is that you want the chair and/or couches to be comfortable. Usually, something that you would have in your living room or family room will do well in your home theater. Also, stuffed chairs and couches will aid the acoustics by catching sounds that might otherwise echo through the room.

Visuals

Not all of your time in your home theater will be spent watching TV. You might just run the audio and listen to a CD. For those times, you need some pleasing visuals on the walls. These typically can be paintings, curtains, or even colorful blankets.

Avoid anything in a striking color that will detract from the TV when it's turned on. To be sure of not causing a visual disturbance, keep all visuals off the front wall where the TV is located. Also, check into colors discussed shortly.

Some people like to have lava lamps or static-discharge lamps, which give color, light, and action. These are nice, if you like them, when you're listening to music alone. Quite frankly, they're extremely distracting when you have your TV on. If you want to use these, be sure you can easily access their on/off switch so you can turn them off when you're watching a movie.

Color

Color is a troubling subject for home theaters. Most people don't like what the experts recommend. In fact, most people go ahead and do whatever they want in terms of color, often choosing the more popular colors available at the time they create their room environment. (Note: lighting is discussed in detail in Chapter 4, "The Big Screen," in the section titled "Getting The Right TV/Light Environment.")

I have no problem with wild colors, as long as you understand the preferred way of doing this. Once you know, if you choose to go your own way, by all means don't let me give you a guilty conscience.

Neutral, Neutral, Neutral

It only makes sense. If you want to draw attention to something, paint it a vivid color like fire-engine red or aquamarine blue. On the other hand, if you want to

keep attention from something, use a neutral color, one that does not attract attention.

In a home-theater setting you want to see the TV, not the wall behind it (although, in your field of vision, the wall behind it may account for as much as 75 percent of what you see with direct-view sets). Therefore, the wall should be painted a neutral color. What's a neutral color? There's only one—gray.

That's right: ideally, the wall behind your TV and front speakers should be painted gray. In the past, TV studios were all painted gray. (Today, of course, elaborate sets fill the studios with all the colors of the rainbow. Pundits claim that is to make up for the less-than-colorful personalities who reside in front of the camera.)

What Is Gray?

Gray is gray, isn't it? Not really. Gray is actually the absence of color. It can be an intense gray to a light gray all the way up to a dull white, which is actually a blending of all colors.

What's important is that the paint is a flat, matte. You don't want gloss, semi-gloss, or even "satin" finish. You want flat, plain, no-glare paint.

If you're going to have curtains, blinds, pictures, or other visuals on the same wall as your system, I suggest you get them in gray as well. You want them to be visible, but not obtrusive—and gray is the color that will accomplish this.

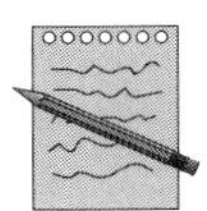

Note: The walls don't all have to be the same shade of gray. Some very clever interior decorating can be done by mixing different shades of gray. An off-white wall with gray drapes and darker grade blinds can be quite pleasant to look at, yet also quite neutral.

The Effects of Color

The problem with using gray behind the television set (and illuminating it properly as indicated in Chapter 4, "The Big Screen") is that, for a while, everything on the television set will look off and you'll be constantly at the color controls trying to change it.

Don't be alarmed. Chances are there's nothing wrong with either your eyes or the TV. The problem is that in the past you were undoubtedly in a room with all kinds of colors, and these affected what you saw on TV. Now, suddenly, you're in a neutral room and the true colors of the TV are coming out and... you don't like them! They don't look natural.

To be sure you're getting the correct TV colors, you can use an SMPTE color setup tape (described in Chapter 4). Or you can simply fiddle with the colors on the television until the skin tones in the picture seem okay. Remember, however, when you first try out your "gray room" it may take you some time to get used to it. Don't expect colors to look right until you've been using it for at least a week.

The Overall Appearance

The ideal to aim for in a home-theater environment is the old acronym, "KISS," which stands for "Keep It Simple, Stupid!" This is not meant to insult anyone's interior-design abilities, but rather to emphasize what the goal is in terms of furniture and light.

Ideally, everything in a home theater should aim at enhancing the audio/video experience. In a conventional living room, where people may spend time chatting, the furniture, wall coverings, color schemes, and so on usually are designed to be bright and cheery (in order, I presume, to keep the conversation on an uplifting note). In a family room, warm tones often are used to express "hearth and home;" a fireplace with logs burning cheerfully is considered a "must" by many on a cool day.

In a home theater, however, what you're after is the best sight-and-sound experience you can get. What's uplifting and warming in other situations is distracting here. That's why a typical home theater may actually seem remarkably dull when you walk into it. But dim the lights, turn up the sound, and put on the movie... and suddenly the room is transformed into a cinematic experience.

A Compromise To Live With

In truth, most of us really don't have a room in our house that's used exclusively for our home theater. Rather, we have a mixed use, multipurpose room; it's a home theater *plus* living room or family room or some other-purpose room.

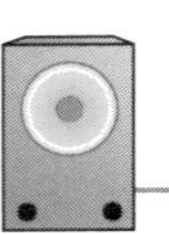

If that's the case, then compromise is in order. Don't paint the whole room gray. Have all kinds of warm colors and furnishings. But at least paint the area behind and around the TV gray so that, when the lights are dimmed, the room's other purposes fade away and the theater experience expands.

You don't have to be a purist. But, it helps to pay a little attention to pure theater needs.

11 CHAPTER

Camcorders and Home Theater

The advent of the modern camcorder has been a real plus for the home-theater enthusiast. Now not only can you show Hollywood flicks, but you can make and show your own high quality home movies. Near-professional quality formats and editing capabilities mean that your homemade sight and sound will look as good—and sometimes better!—than that which you buy or rent at the store.

Integrating a Camcorder into Your Home Theater

A camcorder is not like an A/V receiver/amplifier or a speaker. It doesn't just stay there while you passively sit, watch, and listen. Rather, a camcorder often is a home-theater owner's first introduction into interactive video. With it you become actively involved. You don't just watch the movie—you make it!

Where does a camcorder go in a home-theater setup? Do you put it on top of the TV? Next to the VCR? Or by the speakers?

The answer, of course, is that the camcorder—as the connection between taking and showing movies—often is used when you're catching scenes to show in your home theater, but not when you're sitting there watching them. Unless you're using your camcorder as an A/V playback device (a function all camcorders can perform), chances are you don't need to keep it in your home theater at all (except to show to friends). The essence of the camcorder is to create the tape, which you then play back. It does its job before you roll the movie.

Integrating the camcorder, therefore, mainly means getting a piece of equipment that's comparable in quality to the other components of your system. What that means here is using the superior formats (Hi8 or Super VHS), which record at better than 400 lines of resolution (much better than standard broadcast and VHS tapes) and units that have built-in hi-fi stereo sound. Figure 11.1 shows an S-VHS-C Camcorder.

Figure 11.1. S-VHS-C Camcorder from Panasonic offers up to 400 lines of resolution.

Note: Standard 8mm and VHS only offer about 230 lines of resolution, or roughly what you get from conventional broadcast television.

With high-end camcorders, it doesn't matter what you shoot; whether it be the kids playing at the park or at a soccer game, a party at home, or an actual home movie complete with special effects, the audience will get a theater experience when you play it back. There's nothing to compare with seeing family and people you know up there on your big screen in stereo. It's a part of the home-theater experience you won't want to miss.

What Is a "Camcorder?"

First, let's be sure to understand the equipment. Camcorders, as enthusiasts know, blend a VCR with a video camera, all in a very lightweight, portable body. Today's camcorders typically weigh less than two pounds, even the high-end formats. They allow you to shoot your own video, record it to tape, and play it back. In addition, you can use many modern camcorders as part of an editing system to add titles, special effects, and transitions. (We'll have more to say about this shortly.)

When dealing with camcorders, it's important to remember that the quality of the picture depends on the weakest link. The links in *this* chain include the video camera and VCR inside the camcorder and the playback VCR that you use later on.

The Video Camera

The quality of the image you capture on video tape begins with the pick up, or *eye*, of the camcorder. This image sensor usually is a *charge coupled device* (CCD), a light-sensitive chip. It's rated in *pixels* (short for "picture elements," tiny units of light/size that make up the entire field of a TV screen). The more pixels, the sharper the picture. Modern, high quality camcorders have CCDs with as many as 410,000 pixels. The quality they can produce has been rated as high as 700 lines of resolution or more.

Professional television-broadcasting video cameras use three CCDs for superior resolution and color: one each for red, green, and blue. Until recently, consumer camcorders were limited to one CCD. In 1992, Minolta and Panasonic released two-CCD consumer camcorders. Then, in 1993, both Sony and Panasonic released three-chip consumer models: the VX3 and the AG3, respectively. (Figure 11.2 shows a Sony Handycam PRO VX3.) These are pricey—in the $3,000 plus range—but they allow you to record video with virtually the same quality that can be produced by a TV station. (Indeed, many television stations use these new units for their field work.)

Figure 11.2. The 3-chip Sony Handycam PRO VX3 brings broadcast quality to the consumer.

The Recording Mechanism

Currently there are two competing, high quality video tape recording formats: Hi8 and Super-VHS (known as S-VHS-C in its compact form). The resolution of both is roughly the same: better than 400 horizontal lines. (See Chapter 4, "The Big Screen," for a discussion of horizontal lines of resolution.) However, the Hi8 format tends to have more *drop outs* (tiny flashes or loss of picture pieces) than S-VHS when replayed many times.

In addition, there is a recording-time difference. Currently, Hi8 is available in tapes that run for up to 150 minutes (two and one-half hours). Compact S-VHS tapes are limited to 30 minutes in standard play. (Many camcorders will record up to three times that long in extended or long play. However, the quality is not nearly as good.) S-VHS full size has recording times of 160 or more minutes, although this requires a full-size camcorder. Very few new, full-size S-VHS models are being produced, however, as most people prefer the smaller, lighter-weight, compact units.

The Playback Deck

Every camcorder doubles as its own playback deck. However, many people prefer to have a separate deck for playback built into their home-theater system. This way, they don't have to bother with reconnecting wires every time they want to see what they've shot. (Sony has come out with a Handycam Stand that allows you to simply place the camcorder into the unit to instantly transform it to a playback deck without the need to connect wires. Currently it's only available for the TR-200 and TR-300 models.)

You can buy S-VHS VCRs for a couple of hundred dollars more than standard models. These will play S-VHS-C tapes with the aid of an adaptor. They will not, however, play 8mm tapes. You can now, however, buy 8mm VCRs. These are available from Sony and Samsung. Sony also offers Hi8 playback decks.

When you buy a playback deck, you want to be assured that it also offers stereo sound (assuming that your camcorder has this feature). You also want to be sure that it has certain important editing features, including a jog shuttle and external control, which are discussed shortly (under "Editing").

Features To Look For

Besides looking for a Hi8 or S-VHS-C camcorder with stereo, you'll also want to look for other features that will make your picture taking more enjoyable both for you and your audience. These include built-in digital effects, image stabilization, close-up and zoom lenses, and more.

The idea behind all of the features is to approximate commercial movie making. When you become an videomaker, you enter the heady realm of Spielberg and Lucas. You may or may not be able to duplicate their storytelling abilities, but you can approximate their technical equipment.

Steady Does It

One of the most important considerations with home movie making is to keep the camera steady. Nothing throws the flick off more than to have the picture jiggling on the screen. Put enough inadvertent motion in and you can actually make your audience physically sick!

One answer is the tripod; however, this limits you to a single position. Another answer is "Steadycam Jr." This is a portable balancing device, based on the Steadycam used in the movie industry, which allows you to walk, move, and shoot without jiggling the picture. It costs around $500, and is well worth the expense if you want to shoot professional looking scenes while moving the camcorder.

Another answer is *image stabilization*. This compensates for body movement you may have while shooting. It comes as an added feature in many models. Canon and Sony both use a sensor and a lens-adjusting stabilization feature. Panasonic uses an on-board computer to electronically remove parts of the picture to compensate for unwanted motion. Mitsubishi uses a system of gyroscopes. Each system works differently and has its plusses and minuses. All, however, give you remarkable steadiness. Figure 11.3 shows a Panasonic camcorder that features digital image stabilization.

Figure 11.3. Panasonic camcorder featuring digital image stabilization (DIS).

Through the Lens

Another feature to consider is the close-up/zoom lens. All modern camcorders offer a *zoom lens*, usually with a minimum of 6X. Some offer greater magnification, going as high as 10X or 12X. Yet others, such as Hitachi and Panasonic, connect their on-board computer to the image circuit and achieve digitally enhanced magnifications as high as 100X! These are great when you're taping your kids at a soccer or football game. Figure 11.4 shows the digital zoom control on a camcorder.

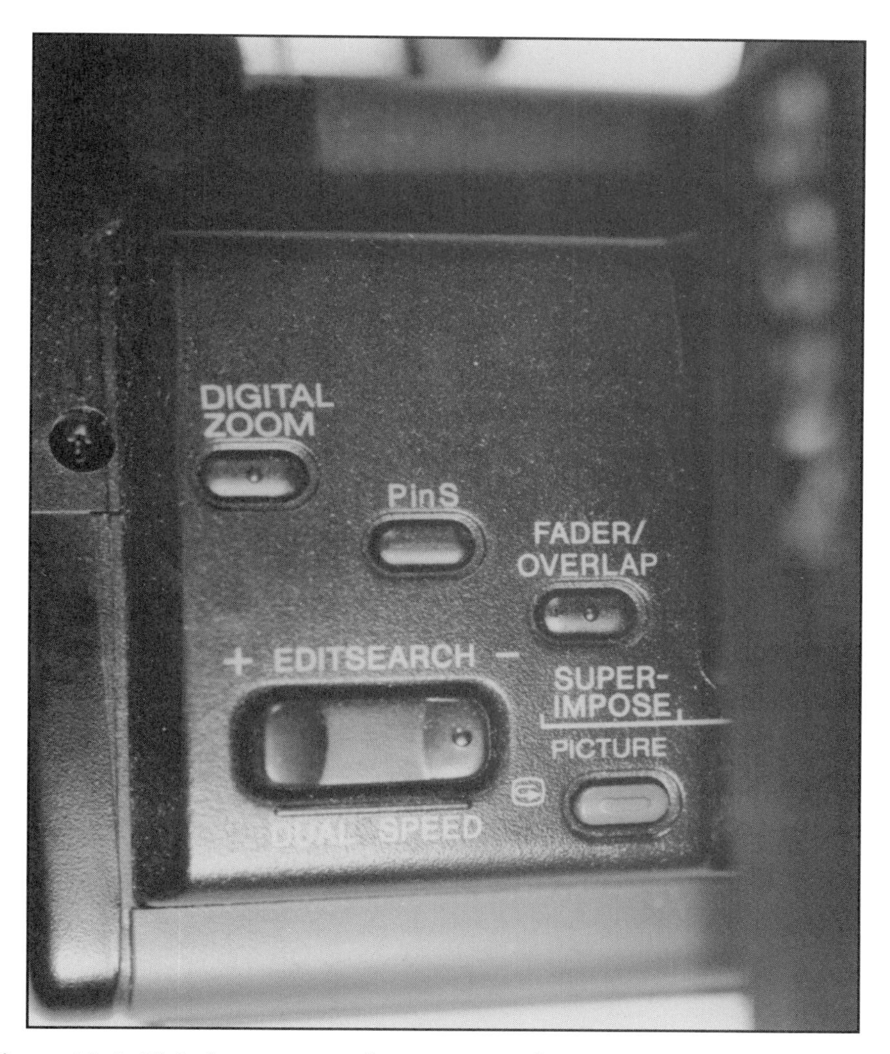

Figure 11.4. Digital zoom control on a camcorder.

Of more importance, however, are *wide-angle lens* features. This allows you to capture the family or a scene without having to back into the next room to get it all in. Wide-angle lenses are now becoming standard on better quality equipment. They generally give you the equivalent of the angle you would get with a 35mm camera.

The Rest of the Menu

Then there are *digital effects*. These include all sorts of fade, strobe, and light effects.

In addition, almost all modern camcorders offer a power-fade control. This is a nice incidental.

Finally, there's a *flying erase head*. This simply means that when you edit scenes together, you don't get a "glitch" or video noise (a bunch of irritating color flashes) between scenes.

Other features to look for in order to take professional quality tapes include manual iris and focus controls, manual white-balance control, and jacks for external mikes and earphones.

What Does It All Cost?

As noted, you can get a three-chip camcorder for around $3,000. You can also, however, get an adequate, one-chip Hi8 or S-VHS-C camcorder, with all of the features mentioned, for about a third as much money. If you don't care about the quality of the picture (but then, why would you be building a home-theater system?), you can settle for standard 8mm or VHS-C for even less. Keep in mind, however, that the camcorder often is only the start. Later you may want to get home editing and desktop-video equipment (discussed later in this chapter), which can cost many thousands more.

Because of the rather substantial cost, a camcorder and related equipment are not usually the first items bought for the home theater. As we'll see in the discussion on interactive video, though, they soon become absolutely vital.

Interactive Video

A home entertainment system, at first glance, would seem to be involved strictly in playback. However, as we've seen, a camcorder allows you take high quality videos for later viewing. In other words, it allows you to participate in the recording as well as the playback.

This "interactive" process, which is just coming on-line, holds enormous promise for the future. (Indeed, some say it's the *only* way into the future!) In the next chapter we'll go into game playing and other interactive activities. Here, however, we're going to consider methods of refining your audio/video interaction to give you even more pleasure.

Editing

Home editing allows you to add titles, fades, wipes, special effects, and more to the videos you've shot. The basic process is quite simple. You play back the tape on one VCR (your camcorder will do). As you record onto another VCR, you add the elements you want while deleting the footage you don't want, all in the order you want it. That's video editing.

The doing, however, is a bit more complex than the telling. A variety of inexpensive and good editing equipment is available. Probably the least expensive is "Thumbs Up" from Videonics, shown in Figure 11.5. It controls your decks through their infrared remote sensors. You simply push an "up" thumb to include a scene or a "down" thumb to discard it. The unit sells for around $200. More elaborate units are available from a host of manufacturers, including Videonics, Sony, Panasonic, Sima, and many more.

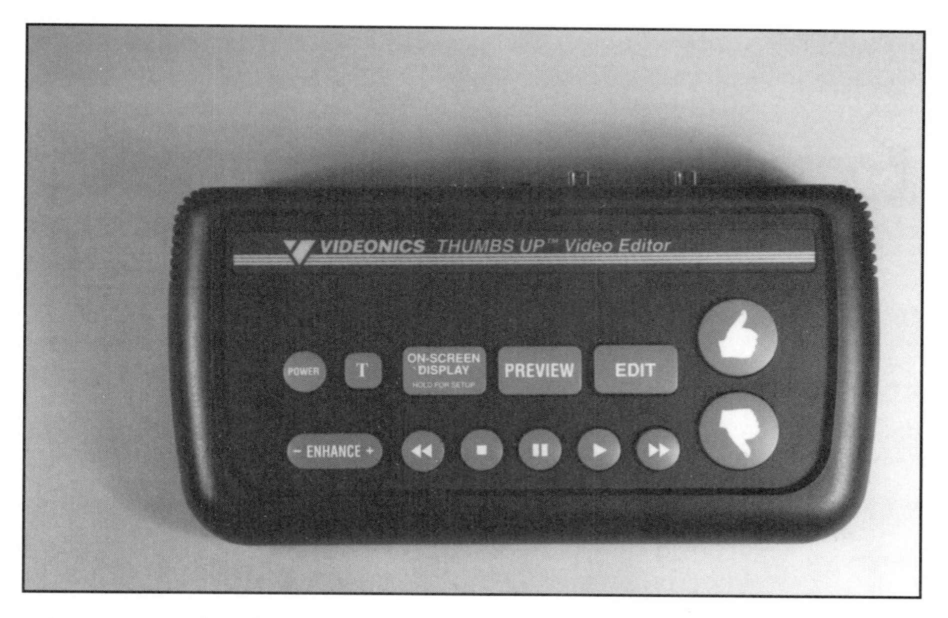

Figure 11.5. "Thumbs Up" editor from Videonics offers quick and easy edit control for around $200.

As you can imagine, what's critical here is lining up the scenes. That's where a jog shuttle on your VCR is important. It allows you to manually forward or advance the tape frame by frame for great accuracy. A built-in flying erase head insures that there will be no glitch (or rainbow of static colors) where scenes blend.

For more professional editing, automatic control of your decks is important. Sony camcorder decks (as well as some Canons and other brands) include a Control-L or LANC (Local Application Network Control) port. Figure 11.6 shows the Sony VDeck with Control-L. Having a LANC port allows you to control the forward, backward, and other functions of your camcorder/deck from your appropriately equipped home computer or editing machine. In other words, you can program the scenes you want in the order you want, and the computer or editing controller properly assembles them.

Figure 11.6. Sony VDeck (foreground) works with computers and camcorders having a LANC port for accurate editing control.

Desktop Video

Of course, once the computer gets involved, the whole world of homemade video changes. Today the trend is to non-linear video (digital video). What that currently means is taking the video you shoot with your camcorder, digitizing it, putting it into your computer, manipulating it, and then sending it out to your record deck. Along the way you can execute a wide variety of effects, including:

- *A/B rolls* (true transitions of once scene to another)
- *Wipes* (where one scene is "wiped" off the screen by a line, color, or some other effect, and another scene is inserted)
- Titles
- Animation

... and more.

The leading desktop video machine as of this writing is NewTek's Video Toaster (shown in Figure 11.7), which runs on the Amiga platform. To use it properly you need the Toaster card and software, the Amiga (currently the 4000), and two time-base correctors (to be sure that the video sources coming in are synchronized). The total cost is around $6,000 to $7,000. FAST electronics has recently introduced the Video Machine, which does many similar things only on IBM clones and Mac machines. The Video Machine costs around $4,000, but you also need the Mac or IBM clone to go with it. Videonics has introduced a digital mixer, shown in Figure 11.8, which does some of the above in a stand-alone unit for only $1,200!

Desktop video equipment lets you create a wide variety of amazing special effects. Remember the movie *Terminator 2*? The liquid creature from the future was created onscreen using *morphing techniques*. These involve "melding" one image with another. You've undoubtedly also seen this in commercials and even in a Michael Jackson music video.

Note: For more information about morphing, see *Morphing Magic*, by Scott Anderson (ISBN: 0-672-30320-5), also from SAMS Publishing.

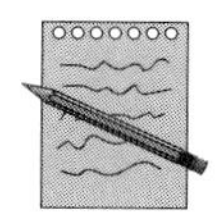

Combining your home computer with your camcorder and associated equipment can very quickly, and at relatively reasonable cost, turn you into a real movie maker.

Figure 11.7. Video Toaster board (and packaging), which revolutionized home video.

Figure 11.8. Videonics' Digital Video Mixer offers wipes, fades, dissolves, and other effects for the amazingly low price of under $1,200.

Interactive Video: The Next Step

Interactive video means that instead of simply sitting back and passively watching a movie in your home theater, you actively participate in the experience. In terms of your camcorder and desktop video, you actually create the movie. Your home theater becomes more than the movie theater, it becomes a playground, a workbench, a learning machine, almost anything you want it to be. To find out what else is available in interactive video, check into the next chapter.

12

CHAPTER

Game Playing at Home

Video game playing is no longer "kids' stuff." It's come of age, with interactive games that provide entertainment and education to all ages. Add a home theater and you've taken game playing to the heights where it becomes almost a total body experience! (It's not quite "virtual reality," but it's close.)

Do you really need a home theater to play games? Of course not. Millions have been playing on Nintendo's GameBoy with a two-inch square screen that's almost invisible, and having a great time at it. Everyone agrees that it's the game itself—the "software," so to speak—that counts the most.

Given good software (in other words, good games), video games are enhanced by being played on a home theater. It's the same old argument about movies. Do you need a home theater to see a movie? Of course not. But what a difference when you do see it on the big home screen with the big home sound.

Traditional Video Game Machines

Nintendo, Sega, and others have been making video game machines for years. Today, the programs are quite sophisticated, and the graphics much improved.

There's no reason, in fact, why you can't take a Sega Genesis machine or a Super Nintendo and plug the video OUTPUT directly into your A/V receiver/amplifier sending the video to your big-screen TV. (Note: you can get stereo out of both Super Nintendo and Sega Genesis except with the latter you need to convert from the headphone jack - phono to RCA adaptor).

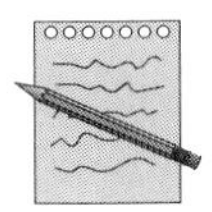

Note: Most video game publishers note that excessive use of their game can potentially "burn" the game's image into your TV screen so that it will be noticeable even when the game is not playing. This was mainly a problem with the early games such as "Tank" or "Pinball" which had a set format. For today's modern games which change scenes rapidly, this is far less of a concern, but nevertheless, you should be aware that the potential problem does exist.

You'll see the game big-screen size and you'll hear the sound amplified. The graphics may look a bit fuzzy in the bigger format (as shown in Figure 12.1). But you definitely will be getting a front-row seat experience.

Home theater is particularly well suited to the many adaptations of games that are coming onto the market. Acclaim Entertainment, for example, makes wireless remote controllers for the Super NES and Genesis video game systems. (See Figure 12.2.) With these you can sit at the back of the room (the game machine up front) and play the game, getting the benefit of big screen/big sound.

Virtual Reality

These days, the talk in the video arcades, indeed amongst home theater aficionados is *virtual reality*. If you don't know what the term means, check out a copy of the film *Lawnmower Man*. It's not a particularly great film, but it does completely explain the concept. For those who don't have access to the film, virtual reality means transporting your body into another world that seems totally real.

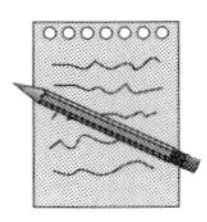

Note: For more information about virtual reality, see *Virtual Reality and the Exploration of Cyberspace*, by Frances Hamit (ISBN: 0-672-30361-2), from SAMS Publishing.

Figure 12.1. A Nintendo game—"NHL Stanley Cup"—on a big-screen TV. ©1993 Nintendo. (Courtesy of Nintendo)

Sound like science fiction/fantasy kind of stuff? Rest assured it's real, and technology is fast making it adaptable to the home theater.

The first true virtual reality machines probably were the flight simulators used by the military and commercial aviation to train pilots. The "player" sat in what seemed to all intents and purposes an actual cockpit. However, instead of windows looking out from the supposed aircraft, there were several television monitors. The entire cockpit, further, was placed on a completely moveable (up/down/sideways) computer-controlled platform. Thus, when the player—the pilot-in-training—banked the aircraft right, the whole cockpit leaned to the right. Inside, the effect was almost indistinguishable from actually flying a plane, and hence its value in training pilots.

Flight simulator games have long been available at home on computers. However, they lack the physical sensation of movement provided by an actual simulator machine.

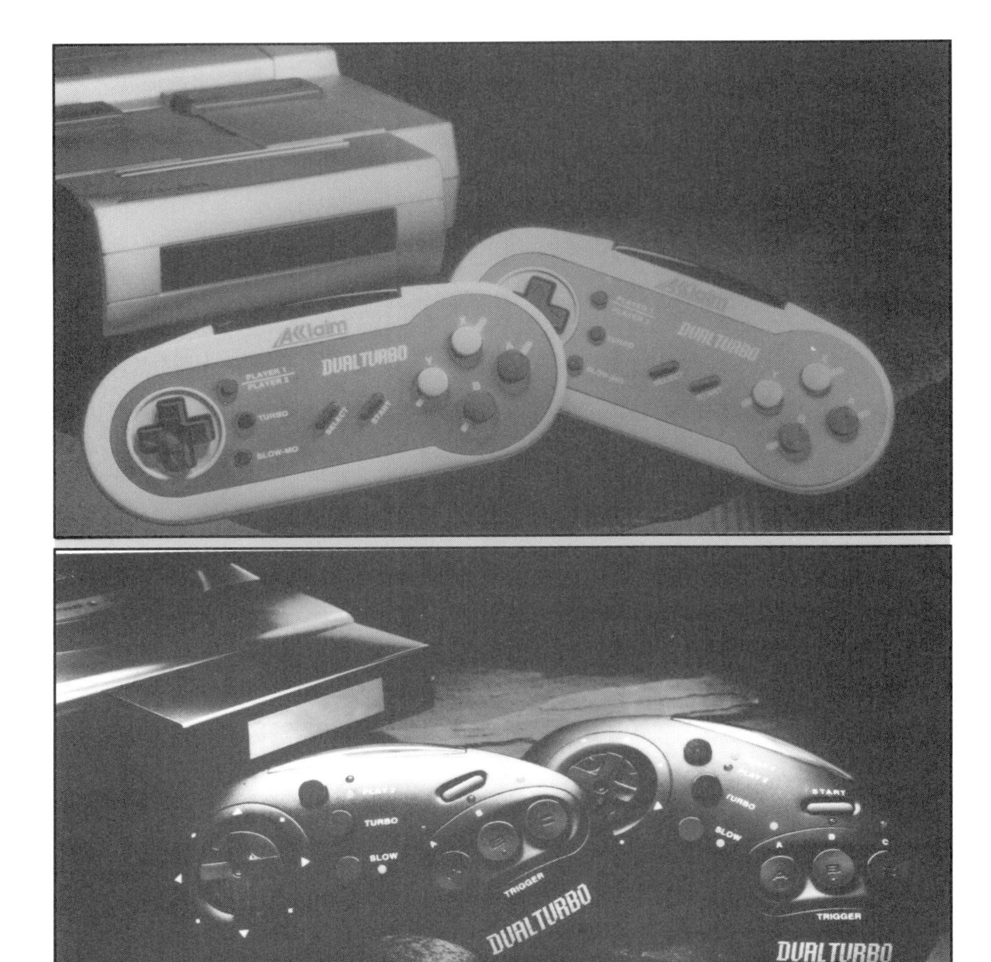

Figure 12.2. Acclaim's DualTurbo wireless remote controllers for Nintendo and Sega.

The game people aren't transporting huge flight simulators into your house. They are coming up with the next best thing: they're producing headsets and earphones that give you an audio/video experience that really is out of this world.

One of the first of these, a Sega VR headset available for the Genesis system, costs under $200 and works in full color and sound. An intuitive control interface enhances the gameplay with sights and sounds coming in from all vectors. Further,

the visual display is 3-D stereoscopic, so it really doesn't let you fall back into your more comfortable reality.

Whether playing *Nuclear Rush* (in which you pilot a hovercraft past intensely hostile robots) or *Outlaw Racing* (where you battle 20 other cars in a careening high-speed race), you're *there*, physically as well as mentally.

It's not just for the kids. Connect one up to your home theater and you'll see what I mean. It's positively infectious! However, a word of warning: These early virtual-reality games are so intense that some people playing them continuously over a long period of time have reportedly gotten physically sick! Best take a break once in a while. (One of the manufacturers even puts out a warning to people with seizure disorders regarding the flicker produced by some of these games. Be careful!)

> **Note:** Virtual reality games are in their infancy. Today there are headsets and earphones. Tomorrow there will be total body outfits that will create another world so powerfully real that you may not be able to tell it apart (at least for short periods of time) from the world in which you currently live.

Interactive Machines

There's a whole new category of machines that are just now appearing and which will allow you to interact with your home theater. Instead of just sitting there and watching and listening, now you can actually give some feedback.

Initially, most of the interaction is in the form of games. However, the sky's the limit on what these various machines can do. We'll consider two of the most likely up-and-comers as of this writing: CD-I from Philips, and 3DO from just about everybody. (This latter will be explained shortly.)

CD-I

Philips introduced its CD-I player in 1991 at about the same time as Commodore introduced its CDTV player. There was much press and hoopla and the response from the public was almost deafening silence.

"CD what?"

"What does it do? How can I use it? Why should I spend $800 for something you can't even describe?!"

Commodore's CDTV hasn't been seen around too much lately. But Philips' CD-I, backed by a reported $500 million budget, is still in the swim of things. Nearly a year after it was introduced, sales of the unit were being recorded in hundreds instead of thousands, but Philips apparently intends to hang in there. At last report, CD-I was being promoted in all of the usual A/V outlets, as well as computer stores and some department stores. In fact, as soon as someone figures out a way to tell the public what this little gem is, sales could skyrocket.

Fore Compact Disc-Interactive!

The CD-I machine allows you interact with a compact disc. The first "game" I saw on this machine, several years ago, involved golf. You were playing golf. You lined up at the tee, prepared your shot, and swung.

Instead of seeing a graphic of a golf ball flying through the air, you saw a video of a real ball on a real golf course. The ball flew through the air and landed, according to how you hit it, at a new location on the course. You marched over to it and prepared to hit it again.

In a sense it was kind of a virtual reality of golf. When shown on the big TV screen (this was one game that really had to be seen big), it really was just like being there. And although I'm no pro at golf, I've been told by those who are that the machine played a pretty good game as well.

Trouble was, the machine cost around $800. For a fraction of that price you could buy a pretty darn good golf game for your PC.

Other Applications

Another early application of CD-I was Compton's Encyclopedia. The entire encyclopedia was on compact disc and played on the machine. And, it was interactive—sort of. You could look up a speech by John Kennedy and see a small video window display Kennedy giving the speech. It was the first truly A/V encyclopedia. Trouble was, again, it just wasn't enough of an incentive to go out and buy the machine.

Philips is now introducing a whole new series of interactive games for CD-I. These include *Caesar's World Of Boxing*, shot live at Caesar's palace in Las Vegas. You can select between an arcade version and a strategy version. Another selection is *Kathy*

Smith's Personal Trainer, which allows you to select goals, music tracks, watch an elapsed-time clock, and get other helpful information while exercising. Figure 12.3 shows Philips' CD-I system.

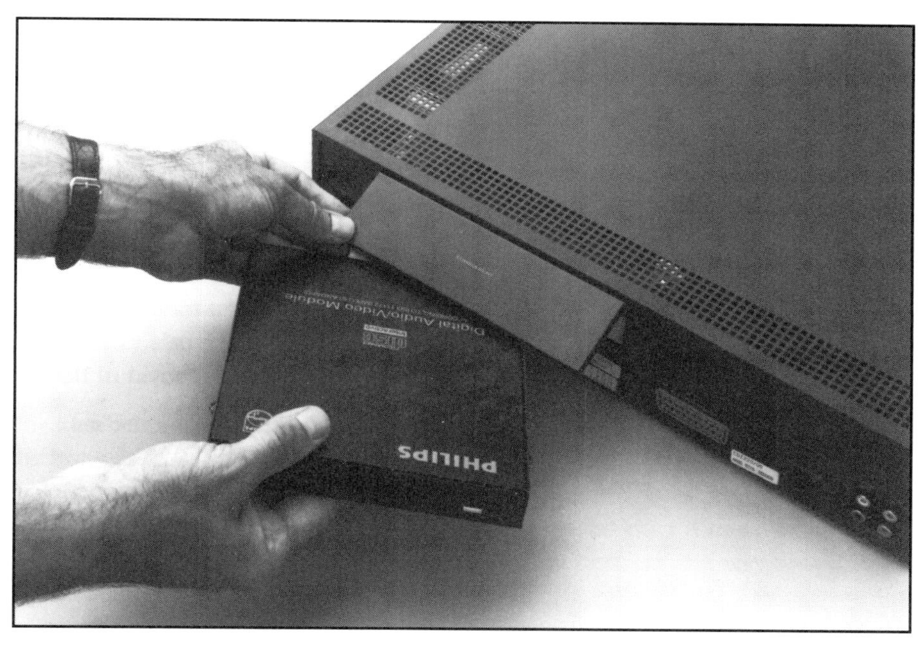

Figure 12.3. Inserting Philips' Full MotionVideo Cartridge for the CD-I system. It offers full-screen video.

Further, arcade games that include live-action video plus unique digital effects are being brought on line. *Star Wars*, based on the George Lucas film featuring original film clips, is one. *Link: The Faces Of Evil*, featuring the Nintendo characters Link and Zelda, is another.

Also, in 1993 Philips signed an agreement with Paramount to put theatrical-length movies on its CD-I system. The movies will be digitized on 5-inch discs.

Unlike a laserdisc, there really isn't room for a whole movie on one CD. In fact, even using MPEG (Motion Pictures Experts Group) video compression, there's still only room enough for 72 minutes of video. Hence, you'll probably need two discs to run an entire film. In addition, you'll need a special, plug-in adaptor that costs around $250. While the total number of movies to be made available isn't known, it's fair to say the numbers will be large.

When I had a chance to view these new CD movies, I found them to be of good quality, but not quite up to the quality of a laser disc. Nevertheless, if you're considering getting just one new machine, the CD-I would be a good bet for versatility.

Perhaps a more exciting development is the addition of music video titles including Sting's *Ten Summoner's Tales* and Tina Turner's *Live In Rio 1988*. Soon, you may be able to watch digital movies and music videos, as well as play games—all on the same machine.

Enter 3DO

3DO is a *multiplayer*: an elaborate home multimedia machine created by the 3DO Company. As mentioned in Chapter 1, the 3DO Company doesn't make anything. Instead, it licenses the technology of its product to other manufacturers. The first out with a player was Panasonic with its R.E.A.L. 3DO Interactive Multiplayer, shown in Figure 12.4.

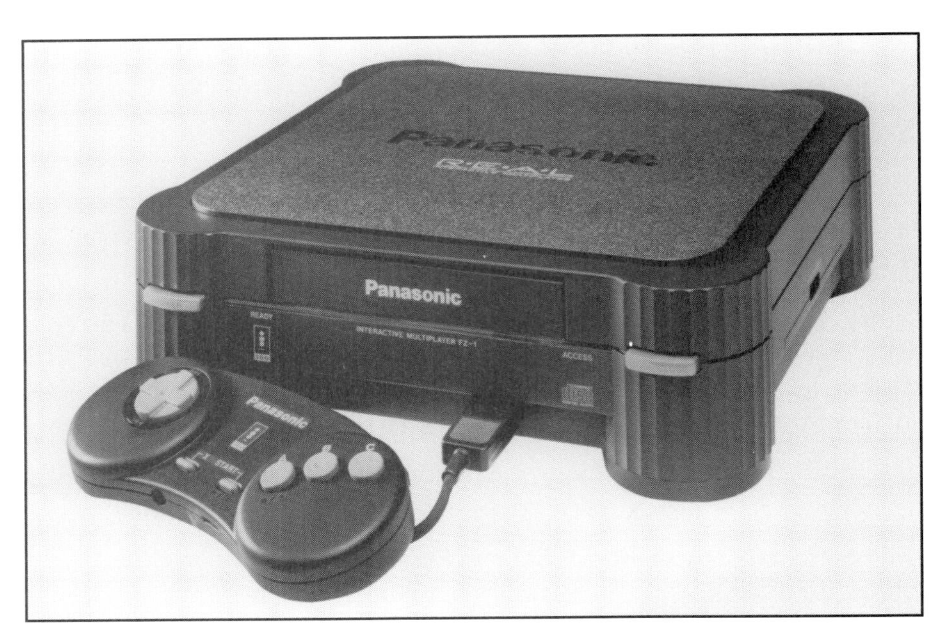

Figure 12.4. Panasonic Multiplayer 3DO, which provides special effects and other features right on your television set.

But, you may be asking yourself, just what does a multiplayer *do*? Let's consider that.

Using CD-ROM technology (that's a compact disc that can be read, but not written to), the machine will play music CDs. It also can display video CDs on your television set. It will have zoom features that allow you to magnify and focus on a particular part of the picture. And, like CD-I, it can access encyclopedias and atlases for instant access to comprehensive information.

With your home theater, you will be able to use it as a special effects generator. Because it has Y/C in and out on the back, the video signal is broadcast quality. ("Y" stands for luminance and "C" for chrominance in the incoming broadcast signal.) You'll be able to divide your television screen into four separate quadrants, each carrying an identical image. Or you can divide the screen in half, with the right side being a mirror image of the left. Using the zoom divide, you can create weird distortions, including warping the picture, rearranging proportions, and spinning the picture around on an axis. You can even layer hundreds of images on the screen at once, animating them together or individually.

Best of all, it happens in real time. That means that as the signal is input from your camcorder, VCR, or broadcast, you can manipulate it while you watch the result on your television set.

And, as you create artistic effects, perform Hollywood-style movie tricks, and even add animation to what you've created, you can record it all on your VCR. Thus, you can take existing movies (either from another VCR or your camcorder) and manipulate them to create a whole new version that's just the way you like it.

Some of the possible effects that you will be able to do (some of which may require additional software development by either 3DO or its licensees) include:

- *Texture mapping*—create a texture for a three-dimensional object. Make it look like wood, or steel, or paper.
- *Pinpoint lighting*—You'll be able to define a light source's location and use it to"illuminate" an object on your screen so that shadows are formed in the appropriate places.
- *Create transparencies*—Effects such as fog, clouds, water, and clear glass can be added to your video.
- *Warp*—You can bend, twist, stretch, or shrink objects that are on your screen. For example, if there's a face on your screen, you can indicate you want to push on the cheek. When you do, it bends inward in a natural fashion.

This is more than just games and interactive video. It's a chance to create unique videos at home. In effect, if you don't like the way the movie looks, you can make it look different!

Initially, 3DO players will likely be more novelty items than anything else. However, some in the industry believe that someday these devices may actually replace VCRs. It could happen sooner than any of us think.

CHAPTER 13

Hooking Up to a Satellite

If you're considering a dream home-theater system, you've got to look into a home-satellite hookup. You don't need satellite to see movies played off laser discs or VCRs. But if you want the ultimate in broadcast TV, satellite is the only answer.

A satellite system allows you to receive broadcast TV with amazing clarity, as much as 400 lines of resolution or more. It also provides excellent stereo and, in many cases, Dolby Surround Sound. Everything you can't get off your cable or antenna, you can get from a satellite system.

Of course, a satellite system does add to your home-theater expense. A good system, with satellite dish and other components can cost around $2,500 installed. And then there's the problem of finding a location for the "dish." We'll get to these problems in a moment, but first, let's be sure we know what's involved.

What Is a Satellite System?

The TV signal that you receive (with the exception of some local stations) is transmitted via satellite. That means that from the broadcast studio, an *uplink* sends the TV signal to a *geostationary* satellite (that is, a satellite at such a height and

traveling at such a speed and direction as to remain in the same location relative to the rotating planet).

The satellite (called a "bird" by those who have home systems) receives the signal and rebroadcasts it over a large expanse of the planet, called its *footprint*. If you're in the satellite's footprint, you can receive the signal. (Of course, you must have a satellite receiving system.)

This whole process was originally described by author/scientist Arthur C. Clarke (*2001: A Space Odyssey* and other works). He theorized that a satellite launched from earth and matching the earth's speed would appear stationary over the equator at a height of roughly 26,000 miles. In other words, by matching the planet's speed, it would appear not to move. Such a satellite would be in a perfect position to receive and retransmit broadcast signals. Today, dozens of such satellites hang in a narrow equatorial band. In the United States, transmissions from satellites located from 137 degrees west longitude (satellite F1) to 69 degrees west longitude (satellite S2) can be received. This amounts to hundreds of different channels.

Currently, all of the *network feeds* (the original broadcasts) are on satellite. All of the premium channels (HBO, Cinemax, Disney, Showtime, and so on) are on satellite. All of the pay channels (ESPN, CNN, and so on) are on satellite. Even the Playboy Channel, Home Shopping Network, and dozens of tiny-but-popular independent channels are on satellite. In short, almost everything you see (except for your local television, which is broadcast line-of-sight) is on satellite.

And, as noted, once unscrambled, the signal is pure and clear. It's possible to get Y/C separation ("Y" for black and white and "C" for color) of the signal for hookup to a high-resolution monitor. Stereo and surround often are available. Thus, to get the best broadcast video, you really do need a satellite system.

Think of it this way: your cable company is just in the way of you receiving the best that's out there. With a satellite, you bypass cable. Best of all, the cost for satellite broadcasts of premium channels is usually about half the cost you would pay to a cable company!

The Elements of a Satellite System

In order to receive home-satellite transmissions you need at least four components:

Dish—A collecting device to receive the signal

Feedhorn—An electronic device that receives the focused signal and transforms it into impulses that a receiver can use

Actuator—A device to move and aim the dish at different satellites
Receiver/descrambler—A device that amplifies, modifies and decodes the signal so that you can see and hear it on your television set

These components are discussed in the next sections. (Of course, there's also all the wiring and installation equipment that go along.)

The Dish

When I first became involved in the satellite business nearly 10 years ago (editing a magazine called *Home Satellite TV*), dishes were huge, cumbersome pieces of equipment. I can recall spending several days putting one up, and then spending another day trying to get the darn thing correctly aimed.

Today it's a different, easier, smaller world of dishes. Today's dishes are quick to set up, and their design is held to strict tolerances so their signal reception is optimized.

Dish Configuration

A satellite dish is essentially a parabolic reflector. The signal coming from the satellite, located tens of thousands of miles out in space is very weak. It generally is under 16 watts *at the source*. That means that signal strength is almost unmeasurable by the time it reaches your home.

To simply try to receive such a weak signal unaided would be beyond the ability of current electronics. What's needed is to increase the signal strength. This is done through the use of a dish. The signal falls over the entire surface of the dish. Because the dish is parabolic and reflective, the signal is focused and reflected back toward a central point where, collectively, it is much stronger.

Dish Materials

Modern dishes often are made of steel mesh or perforated plates. The holes allow you to see through the dish and allow wind to blow through it, meaning it's less likely to be blown over. The dish still reflects enough of the signal to be effective.

Solid reflector dishes of spun aluminum also are used. However, they obscure the view and often require a sturdier support system as they have to absorb all the wind that hits them.

It's interesting to note that in snow country, the solid dishes often fare better. During heavy winters, solid dishes tend to collect less snow than mesh, mesh dishes allow the snow to melt through them, and thus, snow becomes more firmly attached.

Dish Sizes

Dish sizes vary according to how far away from the equator you're located. In the central United States, dishes with diameters of around six to eight feet are generally recommended. That size increases to eight to 10 feet at the coast and 12 feet farther north.

Installations

The biggest concern for most of us who want a home-satellite system to go with our home theater is where to put the dish. Adherents notwithstanding, dishes are big and, for most people, not that attractive.

Some rather clever ways of disguising dishes have been devised. These include shaping dishes like patio umbrellas and lowering them below ground level.

Nonetheless, chances are you're going to need a fairly good-sized piece of land with enough clearance to aim the dish at the southern horizon. Backyard installations are the most common.

Problems you may run into include trees that are in the way. The water in tree leaves tends to block the signal level. With deciduous trees, that means that in the fall and winter, when the leaves are off, you get a strong picture. In the spring and summer, when the leaves are full, the signal is weaker. It's best to avoid trees.

Another problem may be neighbors complaining about the dish. You do have a right to receive satellite signals. However, your property may also have CC&Rs (Conditions, Covenants, and Restrictions) which limit the location of a dish or otherwise restrict its use. Check with your homeowner's association first, if you have a homeowner's association.

Dishes are best set up professionally. The reason is that the post, which usually is cemented into the ground, must be perfectly perpendicular to the earth in order to have the dish rotate properly. Also, the dish itself is cumbersome and difficult to assemble if you're unfamiliar with the procedure. A good crew, however, can have your dish up and running in just a few hours and usually for only a couple of hundred dollars in labor. Dishes themselves cost anywhere from $500 to $1,000.

Future Dishes

The big dishes are needed because most satellite broadcasts are currently on C-band, or 3.72 to 4.18 GHz (channels 1 to 24), which requires a large collecting area. Newer satellites tend to transmit on Ku-band or 11.7 to 12.2 GHz. Dishes needed to

receive this signal need only be only a couple of feet across. Many modern dishes have dual-feedhorns, receiving both C- and Ku-band.

Plans currently are underway for a new satellite system that would offer up to 200 movies on a "pay-per-view" basis using the Ku-band and require dishes no more than 18 inches across! This system pioneered by leading corporations including Hughes, RCA, Thomson, and others may be in operation by 1994.

The Feed

While the dish is certainly the single most obvious part of a home-satellite system, as noted earlier, it's only one of several components. The *feed* is another. This is composed of a *feedhorn*, which collects the signal reflected to the center by the dish, and a *low noise block* (LNB) *converter*, which converts the signal down to lower frequencies that can be used by the receiver. If you're set up for both Ku- and C-band, you'll have two collectors. (When purchasing an LNB, remember that you want the lowest noise temperature (static interference) possible, preferably under 40 degrees Kelvin.)

A coaxial cable (that is, a wire like that used for a cable TV hookup) comes from the feedhorn and runs from the dish into your house, and then to your satellite receiver.

The Actuator

The *actuator* is a device that moves the dish along a preset track so that it can pick up satellites all along the horizon. Many actuators these days are *linear*, meaning that they include an arm that elongates or contracts by a small motor rotating the dish. In the past, circular cog drives were used to actuate dishes. These generally proved to be unpopular.

The actuator must be compatible with the receiver; otherwise, you won't be able to turn the dish from inside the house. (In the old days, in order to move the dish, you had to go out into the yard and crank the actuator by hand!)

The Receiver/Descrambler

Called an *integrated receiver/descrambler* (IRD), these devices take the signal from the feedhorn and translate it into signals that your TV can use. IRDs also include a built-in VideoCipher II Plus descrambler (the most recently introduced one) that allows you to receive premium and pay channels.

There are several well-known manufacturers of IRDs, including General Instruments (which makes the descrambler for almost everyone else), Houston Tracker, Toshiba, and Chaparral. (Figure 13.1 shows the Toshiba Home Satellite Theater.) Almost all IRDs include presets for channels and satellites, as well as full remote control. Some IRDs come with UHF remotes instead of infrared. This means that you don't have to be in the same room with the IRD in order to change the satellite *or* the channel!

Figure 13.1. Integrated receiver/descrambler (IRD) from Toshiba with illustrated horizon dial.

Modern IRDs also offer *automatic polarity control*. Satellite signals are beamed to earth in both vertical or horizontal polarity (in order to squeeze more channels on per satellite signal). In the old days, you had to change polarity yourself. Today, the IRD handles it all for you.

When looking for an IRD for a home-theater system, you'll want to check to be sure that, in addition to excellent video (including a *video noise reduction circuit*, or VNR), it also includes such features as Y/C (separate luminance and chrominance) and surround sound.

A Word About Scrambling

As suggested at the beginning of this chapter, in the early days, people bought satellite systems in order to get free TV. Everything that you could get on cable was available on satellite (because the cable companies were getting their programming off the same satellites).

That was a long time ago, however, and today much of what you will want (including the premium channels and pay-per-view shows) are scrambled. The scrambling is very sophisticated and you won't be able to get the programming without subscribing. (Subscriptions are available when you buy your IRD and, as noted earlier, you generally get the same programming for about half the price of cable!)

There are promoters, however, who will tell you that you can circumvent the scrambling and get free TV off satellite. Indeed, it has been possible to buy black market "black boxes" that decode these signals without having to subscribe. However, scrambling procedures are becoming ever more complex and difficult to crack—and taking the scrambled signal without paying for it *is* a crime. It's simply easier on the conscience and peace of mind to pay for programming.

Furthermore, the reason that most people today have dishes (except for those in rural areas who can't get cable) is to get the highest-quality transmission available. If you go to the trouble and expense of building a home theater, it hardly pays to worry about the cost of descrambling the signal. (It's kind of like the old theory that if you can afford a Cadillac, you shouldn't be worrying about the gas mileage.)

How to Get Started

There are number of publications (including *ON-SAT*, *ORBIT*, and *Satellite TV Week*) that carry listings for programming and also carry advertisements for satellite dealers. Satellite dealers also are found in your phone book. In addition, there's a national association, the Satellite Broadcasting and Communications Association (SBCA), that can provide some information. It can be reached at 703/ 549-6990.

14 CHAPTER

Coming Soon to Your Home Theater

The world of home audio and video is constantly changing. What we've seen in this book is only the beginning of what will be available to us at home over the next decade. In this chapter, we're going to look at some things we can expect to see, based on what's just now coming online.

Full Home Surround

The basic concept in this book is that we'll have a room set aside in our home that will be used for home theater. What that implies is that all the rest of the rooms in the house will not have A/V. In the future, this may change, and significantly.

Back in the 1950s and early 1960s, a concept was introduced called *home communications*. It was built into more-expensive homes, and consisted of a speaker (usually of not-wonderful quality) built into the ceiling of every room. (It was popularly called an intercom system.) There was a microphone set up near the front door, and a control panel that usually was placed in the kitchen. The control panel allowed the user to turn on or off each speaker/room individually, talk to someone who might be at the front door, or tune into AM/FM and play this in any location throughout the house.

The idea was a good one, although it fizzled—probably because builders weren't willing to spend the bucks necessary to use really good equipment and get really good sound. When "home-entertainment" systems came into vogue some time later, with everyone buying their own higher-quality stereo, the idea of "home communications" faded away.

However, it's returning today in some expensive homes, as well as in the homes of many "do it yourselfers." Once again, a speaker is built into the wall of selected rooms (typically bedrooms but, in some cases, even bathrooms!). Figure 14.1 shows a wall-mount speaker used in such a system.

Figure 14.1. The HT3 in-wall speaker from McIntosh Laboratory, Inc. This speaker gives you high-quality sound plus minimal visual disturbance.

Today, however, the speakers being used are of higher quality. Further, satellite speakers are being installed so that you can get stereo in selected rooms. In some cases a television set is installed in the wall and a complete surround system is offered.

What's new here is that instead of using a separate A/V receiver/amplifier for each room (with a separate VCR, separate laser disc player, and so on), there's one central location, typically the home-theater area, where these central components are kept. Only the signal to the TVs and the speakers is sent out to the selected rooms. Thus, instead of a home theater you get, in essence, a theater home!

The key to this system (besides having the money for the extra TVs, speakers, and wiring) is a good switcher and a UHF remote. The switcher allows the homeowner to select which room to pump the A/V into. The UHF remote allows the user to activate the switcher from anywhere in the house. (UHF remote isn't dependent on line-of-sight, as is infrared remote.)

If this idea intrigues you, you can easily build a home theater by expanding the equipment and ideas found in this book.

Still Video

The video emphasis thus far has been almost entirely on video in motion: movies on tape or laserdisc. There is, however, an entire other consideration: *still video.*

Almost everyone has family albums. Most people take photographic pictures using 35mm cameras. The vast majority of people don't see any connection between video and still imagery. It's as if they were in two different worlds.

The truth, however, is that still video is rapidly merging with motion video. For example, Canon (XAPSHOT) and Sony (Mavica) have introduced still-video cameras. Unlike a camcorder, these still-video machines take single frames and store them on a disc. Later, they can be played back directly onto a big-screen TV. In some cases, it's even possible to get a short burst of not-very-clear sound to go with them, so a description of what the shot is can be given.

Going in a different direction, in 1992, Kodak introduced Photo CD, shown in Figure 14.2. This is a combined photographic/video process that has promised to revolutionize the way we see pictures. The revolution has yet to happen, but it's well worth mentioning as an important step toward a comprehensive home A/V system.

In Photo CD, Kodak dealers are set up with a special, computerized system that lets them take your negatives from photographic film shot on a 35mm camera and convert them to digitized files on a compact disc. Then, if you have an appropriate Photo CD player, you can play that disc at home and see your film images on your television set.

The Photo CD holds up to 100 photos; by taking it back to your Kodak dealer, you can have new images added or have old images replaced. And the quality is extremely high. My own estimates are that the picture quality is as much as 16 times that of a conventional television set, and as much as four times that of the new HDTVs, coming soon (and discussed shortly).

Figure 14.2. Kodak's Photo CD player, which lets you see your 35mm prints on your television set.

What's interesting about the Photo CD concept is that once you have your pictures in digitized form, you can manipulate them in a wide variety of ways. With just a little bit of manipulation, they can be called up as files on your computer. Using a wide variety of paint, animation, morphing, and titling programs, you can change them in almost unimaginable ways. You can distort a relative's nose. You can add titles to a birthday-party shot. You can add animated figures to a vacation scene. You can enhance color and contrast. In short, you can change them to suit almost any whim you may have.

Right now it takes a bit of computer know-how to do all of this. Soon it will be much easier. It's a new blending of computers, photography, and home theater.

Virtual Reality

If there's one buzzword that's captured the imagination of the youngsters (as well as many of us oldsters) in the last year or two, it's *virtual reality*. I was at a recent

Consumer Electronics show in Chicago and nothing got more attention than a virtual-reality headset for use on the Sega system. The line to get a chance to try it was longer than any other line to see any other product at the show. (See a further description of this headset in Chapter 12, "Game Playing at Home.")

However, virtual reality now isn't truly a reality for most of us. It tends to be more hype than anything else. We may indeed have virtual-reality games and devices soon. But "soon" can mean several more years.

Today, however, you can have at least one version of virtual reality that relates to home theater. It's called "Virtual Vision," as shown in Figure 14.3.

The product is called Virtual Vision Sport, and consists of a headset and portable television receiver. When you put the headset on and turn on the system, you see a big-screen color television image floating in space about ten feet in front of you. The headset also provides stereo sound, although you may want to turn that feature off in favor of your home-theater system (described earlier in this chapter).

The idea behind the product is that you can watch TV and do other things around the house, such as make dinner or handle the washing, at the same time. At a price tag of around $900, however, this is not going to be the sort of product that you'll want to buy on a whim. Further, although as a first product it has a great deal of novelty value, it's probably more of a transition item on the way toward true virtual-reality products.

Coming soon, perhaps within the next two or three years, will be virtual-reality home theater. Here, instead of just slipping into your easy chair to watch and listen to movies, you'll slip into your virtual reality environment. Whether it be headsets or a surround TV, your vision will be captured by the system, as will your hearing (and, to some extent, your sense of motion and balance). You actually will be able to enter a separate world of reality created by the programmers—a virtual reality.

The price tag for the first of these home systems will be astronomical. Within a year or so, though, it should come down to earth—or true reality (as opposed to virtual).

Figure 14.3. Virtual Vision Sport headset and TV receiver, which lets you watch a TV image projected in front of you.

High-Definition TV

We've talked elsewhere in this book about the coming of high-definition television. However, let's spend a moment or two here to discuss the implications of HDTV for the home theater, in terms of planning.

If you know that HDTV is coming, should you "waste" money on a big-screen TV now? Or should you wait until the new HDTVs are introduced, probably by 1995?

At this writing, HDTV is sufficiently far into the future that few people are seriously worried on this point. However, a consortium of nine different major corporations involved in the field recently agreed to work together on a single standard for HDTV in the United States. Further, the FCC has indicated it will give its blessings to whatever system the consortium comes up with, as long as it meets FCC guidelines (which, translated, means that it will be compatible with current color TV standards. Nobody in government is willing to risk the ire of the public by saying that the new HDTV should only work on special new sets.)

All of which is to say that, any day now, an announcement could be made stating that an HDTV standard has been decided upon and the new sets are on their way. I have heard experts state that within 24 months of a standard being determined, sets could be at your local stores—probably a lot sooner since the first company out there stands to inherit a lion's share of the market.

So, if you're starting to build a home theater, what should you do? Should you wait? Or should you buy today's existing technology?

As editor of a national newsstand magazine in the consumer electronic's field, this is a question I'm frequently asked. Over the years I've come up with an answer that has weathered the test of time: "Always buy today's latest technology. But never wait for tomorrow's promised technology." That's it. If you think it's not overwhelmingly deep, sorry. It's the best I can do. The point is, I've found that this answer works.

If you buy a big-screen TV today, you should be able to spend years enjoying it. Today's sets, either tube or projection, offer excellent brightness and reasonably good clarity.

When the new HDTV sets come out, chances are they won't revolutionize the market overnight. First off, you can count on them being expensive. The first sets in the marketplace are likely to be in the $5,000 range. It will very likely take a good many years before HDTV sets become as plentiful and as inexpensive as modern-day color TVs. During all that time you will continue to enjoy the big-screen set you buy today.

Eventually, you probably will get an HDTV set. At that time you'll undoubtedly move that old big-screen TV set into another room and continue to enjoy it.

In other words, why miss out on all the fun of having a good video set up in your home theater today, just because you're waiting for tomorrow's technology? There will always be something new on the horizon, and you could wait forever hoping to get the final, ultimate TV design.

Fantasy Products

In addition to those we've discussed, there are a whole raft of products that are in the world of fantasy today. We'll cover two areas here: 3-D and scenticizers.

3-D

Today there are 3-D imaging programs readily available on all computer platforms. However, 3-D doesn't mean that something ends up appearing in true three-dimensional form. Rather, it means that the computer programs add perspective to the image so that, like a painting, it gives the illusion of 3-D. There's nothing in computer imaging today that actually produces real 3-D.

About the closest we've come to 3-D, in fact, have been those 3-D films watched through colored glasses and holograms. Both are pretty poor excuses for the real thing, however.

In a home-theater setting, in theory, you should be able to rent a 3-D movie, get the appropriate green/red or other similar glasses, and see the picture as it was originally produced. Don't count on it, however. Although occasionally resurrected, 3-D has never had a long track record of success.

Holograms, on the other hand, are something else. These use laser light to create images that appear to have true depth. The problem is that there has yet to be a way to commercially create a computer controlled laser system to produce continuous, three-dimensional video. I understand that many people are working it. As far as I have heard, though, it's still in the imagination stage.

Closer at hand is a system being developed by at least two Japanese consumer-electronics companies. This system uses two or more semi-transparent screens in the television tube to produce a very realistic 3-D effect without the viewer having to wear glasses.

The concept is quite simple, although the creation of the hardware is, apparently, quite difficult. Instead of the present television tube that has one viewing surface, the new set would have two more surfaces stacked one after the other with a space, perhaps an inch or so, in between. The video picture would be projected from the back, as with conventional sets. However, part of the picture would be illuminated on the rear-most screen, part on the in-between screen, and part on the front screen. The result would be true depth. You would actually be seeing a true, three-dimensional picture.

Problems yet to be overcome include how to make the back screens illuminate part of the picture yet be transparent to other parts, and how to design a video camera that would take just the right kind of shots.

I had the opportunity to see a very clunky prototype of this system and can report that the concept is solid. It definitely gives the 3-D illusion without the need for glasses. However, the version I saw only had two screens, front and back, and the picture tended to bleed from one to the other. If it's ever perfected (and that could come sooner than any of us suspect), it will revolutionize TV viewing.

Scenticizers

Then there are the other senses. So far the A/V experience hasn't gotten to taste or smell. (That's not technically correct, as movie-theater popcorn is definitely a taste/smell experience we've all come to associate with the cinema.) But the capacity to experience these senses too is being developed.

While in Japan a year or so ago, I was introduced to a *scenticizer*. It blew different scents at you and, because the sense of smell is so powerful when used in conjunction with video images, it produced very powerful emotional responses.

The problem was that the scenticizer used chemicals to produce different odors, some much better than others. It had to be constantly refilled with the odorizers, several of which were fairly expensive.

I suspect it will be awhile before this product is widely introduced to the cinema, and even longer before it reaches home theaters. It's going to have to wait until they figure out a way to digitize odor.

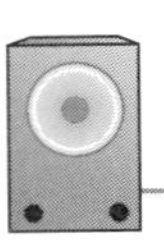

Digital Dolby

Finally, there's a new Dolby system being introduced in some movie theaters. It uses a digital optical soundtrack, provided in addition to an analog track on Dolby Stereo SR D 35mm movie prints. The results are spectacular. The sound is free of distortion, flutter, wow, pops, clicks, and all the other noises that usually occur when you crank up the volume. Further, the dynamic range is similar to what we've come to expect from CDs. It's clear and wide.

The Dolby system offers five full-range channels (left, center, right, left surround, and right surround). It also has an additional bass-effect channel. You may have already heard this in theaters equipped to play Dolby Stereo Digital film. (You may also have heard DTS digital sound used for the movie, *Jurassic Park*.)

Multichannel sound by Dolby and others undoubtedly will grow into the home market, with systems offering greatly increased power and clarity. It's only a matter of time.

If this chapter has tantalized you with what's to come, it's achieved its purpose. Remember, we've only just begun to tap into the future.

Have a Professional Build It

While this book is about showing you how you can build your own home theater, there may come a time when patience ebbs and money (for whatever reason) becomes available. At that time, you may want to give up on the hassle of doing it yourself and have a pro do it. But how do you find a professional at building home theaters?

One answer is CEDIA—the Custom Electronic Design and Installation Association. This is a national trade association of companies that specialize in planning or installing electronic systems for the home. The association was founded in 1989 and has about 300 member companies.

According to CEDIA, its members are established and insured businesses with bona fide qualifications and experience in this specialized field.

To find a CEDIA member in your area or to get further information on the organization, call 800/CEDIA30. Many thanks to CEDIA for photos provided in this book.

Index

Symbols

A

B

D

E

F

G

H

W-Z

Book ISBN 0-672-30381--7